McGRAW-HILL'S

PMP®* Certification

Mathematics

Project Management Professional

EXAM PREPARATION

McGRAW-HILL'S

PMP®* Certification

Mathematics

Project Management Professional

EXAM PREPARATION

Vidya Subramanian, PMP | Ravi Ramachandran, PMP

New York Chicago San Francisco Lisbon London Madrid Mexico City
Milan New Delhi San Juan Seoul Singapore Sydney Toronto

ISBN 978-0-07-163305-5
MHID 0-07-163305-7

Library of Congress Control Number: 2009940167

McGraw-Hill books are available at special quantity discounts to use as premiums and sales
promotions, or for use in corporate training programs. To contact a representative please
e-mail us at bulksales@mcgraw-hill.com.

"PMP" is a registered certification mark of the Project Management Institute, Inc., which was
not involved in the production of, and does not endorse, this publication.

Contents

Acknowledgments   **xi**
Introduction   **xiii**

1 Initiation   1

Introduction 1

1.1 Project-Selection Model 2

   Benefits Measurement 3

   Constrained Optimization 7

   Project-Selection Models Problems on the

   PMP Exam 7

      Sample Solved Problems 8

      Exercise Problems 9

1.2 Accounting Fundamentals for Project-Selection 13

   Benefit-Cost Ratio 13

   Opportunity Cost 13

   Sunk Cost 14

   Depreciation 14

   Return on Sales 16

   Return on Assets 16

   Return on Investment 17

   Working Capital Ratio 17

   Break-Even Analysis 18

   Accounting Fundamentals for Project-Selection

   Problems on the PMP Exam 19

      Sample Solved Problems 20

      Exercise Problems 22

1.3 Make-or-Buy Decision 26

Make-or-Buy Analysis 27

Make-or-Buy Decision Problems on the PMP Exam 28

Sample Solved Problems 28

Exercise Problems 30

2 Planning   39

Introduction 39

2.1 Probability Fundamentals 41

Probability of an Event Occurring 41

Probability of an Event Not Occurring 42

Mutually Exclusive Events 43

Events That Are Not Mutually Exclusive 43

Independent Events 44

Dependent Events 44

Probability Fundamentals Problems on the
PMP Exam 45

Sample Solved Problems 46

Exercise Problems 52

2.2 Expected Monetary Value 56

Expected Monetary Value Problems on the
PMP Exam 57

Sample Solved Problems 58

Exercise Problems 60

2.3 Decision Tree 65

Decision Tree Problems on the PMP Exam 66

Sample Solved Problems 67

Exercise Problems 68

2.4 Project Cost Estimation 76

Analogous Estimating 77

Bottom-up Estimating 77

Parametric Estimating 78

Types of Costs 78

Project Cost Estimation Problems on the PMP Exam 79

    Sample Solved Problems 80

    Exercise Problems 81

2.5 Critical Path and Float 86

    Network Diagramming Methods 87

    Critical Path and Float Calculation Methods: CPM,

    PERT, and GERT 88

    Critical Path and Float Problems on the PMP Exam 89

        Sample Solved Problems 90

        Exercise Problems 94

2.6 Crashing 101

    Fast Tracking 101

    Crashing 102

    Crashing Problems on the PMP Exam 102

        Sample Solved Problems 103

        Exercise Problems 104

2.7 PERT, Standard Deviation, and Variance 114

    PERT, Standard Deviation, and Variance Problems

    on the PMP Exam 115

        Sample Solved Problems 116

        Exercise Problems 118

2.8 Contract Cost Estimation 124

    Fixed Price 124

    Cost Reimbursable 125

    Time and Materials 125

    Contract Cost Estimation Problems on the

    PMP Exam 126

        Sample Solved Problems 127

        Exercise Problems 130

2.9 Point of Total Assumption 138

    Fixed Price Plus Incentive Fee Contract 139

    Cost Reimbursable Contract 139

    Point of Total Assumption Problems on the

    PMP Exam 140

        Sample Solved Problems 141

        Exercise Problems 142

2.10 Determine Communication Channels 147

   Communication Channels Problems on the

     PMP Exam 148

       Sample Solved Problems 149

       Exercise Problems 151

3 Execution   155

Introduction 155

     Earned Value Analysis: Assessment 156

     Earned Value Analysis: Forecast 156

3.1 Earned Value Analysis: Assessment 156

     Planned Value 157

     Earned Value 157

     Actual Cost 157

     Cost Variance 157

     Cost Performance Index 158

     Schedule Variance 158

     Schedule Performance Index 158

     Earned Value Analysis: Assessment Problems

       on the PMP Exam 161

         Sample Solved Problems 162

         Exercise Problems 163

3.2 Earned Value Analysis: Forecast .167

     Budget at Completion 167

     Estimate to Complete 167

     Estimate at Completion 168

     Variance at Completion 169

     Earned Value Analysis: Forecast Problems on

       the PMP Exam 169

         Sample Solved Problems 170

         Exercise Problems 171

4 Monitoring and Controlling 177

Introduction 177

    Probability Distribution 178

    Quality Control Tools 178

4.1 Probability Distribution 178

    Beta Distribution 179

    Triangular Distribution 179

    Uniform Distribution 180

    Normal/Bell Distribution 180

    Probability Distribution Problems on the

    PMP Exam 181

        Sample Solved Problems .182

        Exercise Problems 183

4.2 Quality Control Tools 188

    Control Charts 188

    Bar Charts 190

    Histograms 190

    Pareto Charts 190

    Cause and Effect Diagrams 191

    Run Charts 192

    Scatter Diagrams 193

    Statistical Sampling 193

    Flow Charts 194

    Inspection 194

    Check Sheets 195

    Quality Control Tools Problems on the PMP Exam 195

        Sample Solved Problems .196

        Exercise Problems 197

5 Closing 201

Introduction 201

5.1 Statistical Concepts 201

Measures of Central Tendency: Mean, Median, and Mode 202

Dispersion of the Data: Standard Deviation, Variance, and Range 203

Statistical Concepts Problems on the PMP Exam 206

Sample Solved Problems 207

Exercise Problems 208

A Venn Diagrams 211

B PERT Analysis Using Microsoft Project 213

Index 217

Acknowledgments

This book is dedicated to our adorable newborn daughter Rhea, whose babbles have added a new meaning to life. This book is also dedicated to our wonderful parents whose emphasis on education and commitment has made a difference to our lives. To our sisters Bhanu, Sumathi, Vidya, Subha, and Kavitha who have been a pillar of strength, support, and inspiration to us.

We also want to acknowledge and appreciate Charles Wall of McGraw-Hill for helping us finalize the contents of this book. Finally, we would also like to dedicate this book to all those PMP aspirants who are pursuing the PMP exam in pursuit of excellence in the Project Management field.

Introduction

When we were preparing for our Project Management Professional (PMP) exam, we found that test takers were required to know quite a bit of math to solve the problems on the exam. However, there is no single resource for test-takers that focuses on just the math topics. We wrote this book to bridge that gap for future PMP test-takers.

This book covers all the math concepts that are tested on the PMP exam. Each chapter focuses on one of the five project management process groups, and particularly on the math calculations relevant to that process. Sample solved problems and solved exercises are provided to reinforce the concepts and to show how math formulas are to be applied. The chapters also attempt to give readers valuable problem-solving tips and strategies.

Beyond PMP exam preparation, this book serves another purpose as well. Project Managers who have passed the PMP exam and who have begun their careers can use it as a handy reference guide for solving the math and statistics problems that every project entails. Filled as it is with sample solved problems, this book can help you make informed decisions in all kinds of on-the-job situations.

Under the rules of the Project Management Institute (PMI), we of course cannot share any actual exam questions. However, the questions in this book are modeled closely and accurately on real exam questions. Study them carefully because they will give you a very good idea of what to expect on the test.

WHAT TO EXPECT ON THE PMP EXAM

The PMP exam consists of 200 questions that must be answered in 4 hours. Of those 200 questions, only 175 are graded. The remaining 25 questions are pre-test questions that are being analyzed and validated for use on future exams. These ungraded questions are placed randomly in the exam. You will not be told which questions are not graded.

According to the *PMP Handbook*, the questions are distributed as follows:

Domain	Percentage of Questions
Initiation	11%
Planning	23%
Executing	27%
Monitoring and Controlling	21%
Closing	9%

Out of the 200 questions on the exam, approximately 25 involve mathematical computation. Thus your knowledge of PMP math can have a huge impact on your final score.

HOW TO USE THIS BOOK

This book acts as a guide and supplement to the Project Management Body of Knowledge (PMBOK®) for aspiring PMPs. The purpose of each chapter is to discuss in great detail the math and statistical concepts that are most relevant to a particular PMP process group. Read the instructional material in each chapter, then use the sample solved problems and exercise questions to get a thorough understanding of the different types of problems you can expect on the PMP exam. The goal is to learn the concepts by practice rather than by rote.

The book also provides useful add-ons, including the companion CD that presents additional practice problems covering all topics. The appendix of the book introduces Venn diagrams to Project Managers and discusses the use of Microsoft Project to perform PERT analysis.

FORMULA CHEAT SHEET

Use this handy "cheat sheet" to help you learn and memorize the formulas used most often in PMP math problems.

Project Selection	
Benefit Cost Ratio	Benefits/Cost
Return on Sales (ROS)	(Net income before or after taxes/Total Sales) $\times$ 100%
Return on Assets (ROA)	(Net income before or after taxes/Total Asset Value) $\times$ 100%
Return on Investment (ROI)	(Net income before or after taxes/Total Investment) $\times$ 100%
Net Present Value (NPV)	Project with greater NPV is a better choice.
Present Value (PV)	$FV/(1 + r)^n$
Working Capital	Current Assets – Current Liabilities
Straight-Line Depreciation	Asset Value/Life of Asset
Declining Balance Depreciation	Current Asset Value/Estimated Total Life where: Current Asset Value = Original Asset Value – Previous Depreciation
Expected Monetary Value	
Expected Value	Probability of Outcome $\times$ $ Impact of Outcome
Make or Buy Analysis	Upfront Make Cost + (Maintenance Make Cost $\times$ Z) = Upfront Buy Cost + (Maintenance Buy Cost $\times$ Z) where: Z is the duration at which the cost of making equals the cost of buying, i.e., the break-even point.
Critical Path Method	
Float	LS – ES or LF – EF
Crashing	Crash critical path activities that have a lower crash cost

PERT Analysis w/Beta Distribution	
Expected Duration (E) of an Activity	$[P + (4 \times M) + O]/6$
Activity Standard Deviation	$(P - O)/6$
Activity Variance	$[(P - O)/6]^2$
Project Standard Deviation (Sigma)	Square root of the sum of all activity variances
Project Expected Duration	Sum of all expected activity durations
1 Sigma	68.26%
2 Sigma	95.46%
3 Sigma	99.73%
4 Sigma	99.99%
Three-Point Estimates	

Triangular distribution
    Mean
    Variance
	$(P + M + O)/3$
	$[(O - P)^2 + (M - P) \times (M - O)]/18$

Beta Distribution
    Mean
    Variance
	$(P + 4M + O)/6$
	$[(P - O)/6]^2$

Cost Management and Earned Value Analysis	
Cost Variance (CV)	$EV - AC$
Schedule Variance (SV)	$EV - PV$
Cost Performance Index (CPI)	EV/AC
Schedule Performance Index (SPI)	EV/PV
Estimate at Completion (EAC)	BAC/CPI; used when there are no variances or the spending rate remains the same. $AC + ETC$; used when original estimates were flawed. $AC + (BAC - EV)$; used when variances are atypical of the future. $AC + (BAC - EV)/CPI$; used when variances are typical of the future.

Cost Management and Earned Value Analysis (*Continued*)	
Estimate to Complete (ETC)	EAC − AC
	BAC − EV; used when variances are atypical of the future
	(BAC − EV)/CPI; used when variances are typical of the future
Variance at Completion (VAC)	BAC − EAC
To Complete Performance Index (TCPI)	(BAC − EV)/(BAC − AC)
Rough Order of Magnitude Estimate	−50% to +100%
Definitive Estimate	−5% to +10%
Communication Channels	
Number of communication channels	$[N(N-1)]/2$
Probability Basics	
P (A) and P (A′) are complements	P (A) + P (A′) = 1
P (A or B)	P (A) + P (B) − P (A and B)
P (A and B) for Independent Events	P (A) × P (B)
P (A or B) for Mutually Exclusive Events	P (A) + P (B)
Sigma	
1 Sigma	68.26%
2 Sigma	95.46%
3 Sigma	99.73%
4 Sigma	99.99%

Contract Costs	
Incentive/Bonus to Seller	(Target Cost – Actual Cost) × Seller's % of Cost Savings
Final Contract Cost/Fee or Overhead (given to seller)	Target Fee + Incentive
Total Cost/Price of Procurement	Actual Cost + Final Contract Cost
Point of Total Assumption (PTA)	
PTA for CR contracts	Target Cost + {[Ceiling Price – (Target Cost + Fixed Fee)]/Benefit Share}
PTA for FP contracts	Target Cost + ([Ceiling Price – (Target Price)]/Benefit Share)

CHAPTER 1

Initiation

This chapter covers

1. Project selection
   - Based on benefits measurement
   - Based on constrained optimization
2. Accounting fundamentals for project selection
   - Benefit-cost ratio
   - Opportunity cost
   - Sunk cost
   - Depreciation
   - Return on sales (ROS)
   - Return on assets (ROA)
   - Return on investment (ROI)
   - Working capital ratio
   - Break-even analysis
3. Make-or-buy decision

INTRODUCTION

Project Initiation is the first of the five process groups in the process cycle: initiation, planning, execution, monitoring and controlling, and closing. During initiation, the project is defined through the creation of an approved and authorized Project Charter. The Project Charter officially recognizes the project and identifies the business need for the project and its relevance with regard to the overall strategic plan of the organization. The charter also assigns and entrusts a Project Manager with the responsibility and authority to drive the project to successful completion. Certain components of the initiation process may even be performed outside the boundaries of the project. For instance, many organizations prefer to formally authorize and fund the project even before a Project Manager is assigned to draft the charter and define the scope. Regardless of how the project boundaries are defined, one thing is certain: before embarking on the project, preliminary analysis is needed.

Issues concerning project selection and projected cost savings or returns must be analyzed. This preliminary analysis consists of identifying all the stakeholders in the project and gathering information about what the project requires. Once the preliminary scope is ready, the Project Manager can document the assumptions and constraints that were identified.

In the initiation phase, a Project Manager needs to help senior management decide on the viability of the project. Often, projects are competing for resources and funding, and management must choose the one that provides maximum return to the company. This is where a Project Manager uses math, statistics, and accounting expertise to provide a holistic view of how profitable the venture would be for the organization. In the sections that follow, we will walk through project-selection methods and accounting fundamentals, and we will also review the "make-or-buy" and "buy or lease" analysis.

1.1 PROJECT-SELECTION MODELS

Main Concept

There are two broad categories of project-selection models: the comparative approach (benefits measurement) and the mathematical approach (constrained optimization). Table 1.1 shows some examples of each category:

Table 1.1 Project-Selection Models and Examples

Types of Project-Selection Models	Examples
Benefits measurement (comparative approach)	• Scoring models • Cost-benefit analysis • Review board • Economic models
Constrained optimization (mathematical approach)	• Linear programming • Integer programming • Dynamic programming • Multiobjective programming

We will first review the benefits measurement models and then the constrained optimization models.

Benefits Measurement

ECONOMIC MODELS

- **Payback Period**   The payback period is the amount of time required to recuperate the costs incurred on a project. In accounting terms, it is also referred to as the time it takes to reach the break-even point. If the cost of the project is $200,000 and the revenue for the project is estimated at $50,000 per year, the payback period (break-even point) for that project will be 4 years. Payback period can be expressed as years, months, or even days for small projects. The shorter the payback period, the more lucrative the project will be.
- **Internal Rate of Return (IRR)**   It is the rate at which an investment will yield returns. Normally, IRR is expressed as a percentage. So when the IRR is 5%, it means that for every $100 that you invest, the return on investment will be $105. The higher the IRR, the more lucrative the project will be. The IRR is also referred to as the "return on investment," discussed in more detail in Section 1.2.
- **Present Value**   When an investment is made, it compounds yearly at the IRR for the duration of the investment. Consider a 2-year investment of $100,000 with an expected rate of return of 5%: the yield in the first year is based on the principal investment of $100,000. This initial investment is also termed the "present value" of the investment. The 1-year yield amounts to $105,000 ($100,000 + 5% interest). The yield for any subsequent year is based on the principal investment of $100,000 plus any interest accumulated in the previous year(s). In the case of this example, the principal investment to be used for calculating the yield at the end of 2 years would be $105,000. A 5% rate of return on this compounded principal would yield $110,250. This calculated yield also represents the "future value" of the investment. The formula given below uses the concept of compounding interest discussed above to calculate the future value of an investment.

$$\text{Future value} = \text{present value} \times (1 + \text{interest rate})^{\text{time period}}$$

Using this formula in the example reviewed above, you get

$$\text{Future value} = 100,000 \times (1 + 0.05)^2$$
$$\text{Future value} = \$110,250$$

- **Net Present Value**   In the case of large projects, the company's investment in the project is not a one-time event. Also revenue may be realized over several years. Due to this fluctuation in the inflow and outflow of investment capital, the present value of the project varies over the life of the project. Hence the net present value (NPV) represents an attempt to independently compute the present value for each year and then aggregate all of those values. If the NPV is greater than 0, the project is a lucrative one. If the NPV is equal to 0, the project is neither profitable nor unfavorable. If the NPV is less than 0, the project is unfavorable.

SCORING MODELS

A scoring model is a point-based system using known facts to predict future outcomes. There are different kinds of scoring models used in the industry. The most popular is the FICO (Fair Issac Corporation) credit score model adopted in the United States. The FICO score is used to represent the credit-worthiness of a person and is based on an analysis of that individual's credit and financial history. A high credit score reflects a person's credit worthiness and indicates that the person is not likely to default on loans.

The Project Manager needs to establish a similar scoring model that can be used to objectively examine the "project-worthiness" of a project and compare it to other projects. For example, consider two software development projects: Project A and Project B. Management has set forth parameters for determining the worthiness of each project. Since some parameters are more important than others, weights have been assigned to each one. These weights will be used later in calculating a weighted average project score (see Table 1.2).

Table 1.2 Project Parameters and Weightage		
Scoring Parameter	**Description**	**Weight**
In-house resource availability	This is important to the senior management so that they do not have to incur extra costs in hiring a contractor.	2
Technical know-how	Having the technical expertise to execute the project makes it more favorable.	2
Past experience	Having prior experience with similar projects also makes the project more favorable. Historical information and lessons learned can be applied to increase chances of success.	4
Return on investment	ROI is rated by management as one of the most critical factors in project selection.	5
Intangible benefits	Intangible benefits, such as customer goodwill, may result from the selection and completion of the project.	1
Break-even time	Break-even time is the time required to recover the cost of the project and move into the "black."	3

Table 1.2 Project Parameters and Weightage (*Cont.*)

Scoring Parameter	Description	Weight
Low risk	Management favors projects with lower risks of failure.	4
Maintenance cost	Management favors projects with lower maintenance costs, hence resulting in lower life-cycle costs.	3

Now that the scoring parameters have been defined, you need to choose a scale to use for your scoring system. The scale can be a two-point scale (for example: 1—agree, 0—disagree) or a variation of a polytomous scale (for example: a five point scale is: 1—strongly disagree, 2—disagree, 3—neither agree nor disagree, 4—agree, 5—strongly agree).

To implement the scoring model, you first select a team of experts who will familiarize themselves with both (or all) of the competing projects. Now, assuming that the organization has adopted a five-point scale to choose between two projects, the team has to rate/score the projects. The weighted score for each parameter is calculated as shown below:

$$\text{Parameter weighted score} = (\text{parameter score} \times \text{parameter weight})$$

$$\text{Average project score} = (\text{sum of parameter weighted scores}) \div (\text{sum of weights})$$

Parameter	Project A Score	Project B Score	Weight	Project A Weighted Score	Project B Weighted Score
In-house resource availability	3	2	2	6	4
Technical know-how	4	3	2	8	6
Past experience	5	4	4	20	16
Return on Investment	5	5	5	25	25
Intangible benefits	4	3	1	4	3
Revenue	5	5	3	15	15
Risks	3	3	4	12	12
Maintenance cost	5	2	3	15	6
Total project score				105	87
Average project score				4.375	3.625

In this case, Project A has the higher score and hence would be the choice between the two competing projects.

REVIEW BOARD

Review board is a subjective approach to deciding between competing projects. The review board is composed of numerous experts who bring different perspectives to the table. The board members evaluate the projects based on their knowledge of the industry, market conditions, the company, and the competing needs of internal projects. Two commonly used subjective approaches by the review board are Brainstorming and the Delphi Method.

- **Brainstorming**   This is the process of assembling a diversely experienced team and cultivating an atmosphere that encourages "out-of-the box" thinking. It is a moderated discussion, in which each problem is examined and potential solutions are recorded. These potential solutions are then examined in greater detail to reveal their pros and cons. The objective should always be to foster a healthy debate and come up with recommendations. This technique extends the "two heads are better than one" concept of group thinking.
- **Delphi Method**   This is a way for a panel of experts to weigh in anonymously on the viability of competing projects. A questionnaire might be an effective tool to solicit responses from the experts. The anonymity of the process safeguards the final decision from being influenced by any single member/expert having more clout in strategic decisions. The success of the Delphi method is, however, dependent on the choice of experts and their areas of expertise. Having the wrong set of experts on the panel can produce skewed results and decisions. The expert panel should be representative of the majority of stakeholders identified thus far for the project.

There are no math concepts involved in the review board approach; it is purely subjective. We have included it here for the sake of completeness.

COST-BENEFIT ANALYSIS

In cost-benefit analysis, the potential benefit and cost are estimated based on facts gathered by the Project Manager. The cost and benefit are then compared and expressed as a ratio. If the ratio is more favorable to cost, then the cost of the project is higher than the benefit. If the ratio is more favorable to the benefit, the benefit outweighs the cost of the project. Assume the cost of a project is $400,000 and the benefit is estimated at $800,000. In this case the cost-benefit ratio would be 400,000/800,000. Hence we can conclude that the benefit is more favorable by a ratio of 2/1.

Constrained Optimization

LINEAR PROGRAMMING

Linear programming uses a mathematical model to determine the best outcome, be it maximizing profits or reducing costs. In this process, different outcomes are evaluated by representing certain requirements as linear equations.

INTEGER PROGRAMMING

Integer programming is another way of using math to determine the optimal result. However, it can be used only when there is a discrete set of decisions to be made from a finite set of alternatives.

DYNAMIC PROGRAMMING

Dynamic programming involves using past history to develop a model that depicts the time taken to complete a certain project task. When new values are input, the model generates output that tells you how long a new project is likely to take.

Project-Selection Models Problems on the PMP Exam

TYPES OF PROBLEMS TO EXPECT

There are only five types of problems you can expect in this section of the exam:

(1) Computing the present value
(2) Computing the future value
(3) Choosing between projects given the internal rate of return
(4) Choosing between projects given the net present value
(5) Choosing between projects given the payback period

DIFFICULTY LEVEL

You do not need to have a strong math background to solve project-selection model problems. Once you are through with the exercises in this chapter, you should be ready to face the exam questions. For questions regarding internal rate of return, net present value, and payback period, there are no calculations. You only have to choose between projects. However for calculating present and future values, you have to remember the formulas and compute.

NUMBER OF PROBLEMS TO EXPECT

Expect at least one problem regarding present value and future value on the exam. Also, there will be a problem or two regarding project selection using internal rate of return, net present value, and payback period.

FORMULAS

$$\text{Present value} = \text{future value}/(1 + \text{interest rate})^{\text{time period}}$$

INSIDER TIPS

- When choosing a project based on the internal rate of return, all other criteria being equal, choose the project with the highest internal rate of return.
- When choosing a project based on the payback period, all other criteria being equal, choose the project with the shortest payback period.
- When choosing a project based on the net present value, all other criteria being equal, choose the project with the highest net present value.

SAMPLE SOLVED PROBLEMS

1. A Project Manager wants to invest money into a project and make a return on investment of $110,250 after 2 years at a 5% rate of return. What is the present value of the project?

   To compute the present value, use the formula:

   $$\text{Present value} = \text{future value} \div (1 + \text{interest rate})^{\text{time period}}$$

   Plugging in the numbers,

   $$\text{Present value} = 110{,}250 \div (1 + 0.05)^2$$
   $$\text{Present value} = \$100{,}000$$

2. A Project Manager wants to invest money into a project. What is the future value after 2 years if the present value is $100,000 and the rate of return is 5%?

   To compute the future value, use the formula:

   $$\text{Present value} = \text{future value} \div (1 + \text{interest rate})^{\text{time period}}$$

Plugging in the numbers,

$$100,000 = \text{future value} \div (1 + 0.05)^2$$
$$\text{Future value} = \$110,250$$

3. A Project Manager is trying to decide between Project A, which has a net present value of $45,000, and Project B, which has a net present value of $85,000. All other criteria being equal, which project should she choose?

Remember that when you have to choose between two projects given the net present value, if all other criteria are equal, you always choose the project with the *higher* net present value. In this case, that is Project B.

4. A Project Manager is trying to decide between Project A, which has an internal rate of return of 25%, and Project B, which has an internal rate of return of 15%. All other criteria being equal, which project should he choose?

Remember that when you have to choose between two projects given the internal rate of return, if all other criteria are equal, you always choose the project with the *higher* internal rate of return (IRR). In this case, that is Project A.

5. A Project Manager is trying to decide between Project A, which has a payback period of 2 years, and Project B, which has a payback period of 4 years. All other criteria being equal, which project should she choose?

Remember that when you have to choose between 2 projects given the payback period, if all other criteria are equal, you always choose the project with the *shorter* payback period. In this case, that is Project A.

EXERCISE PROBLEMS

1. A Project Manager wants to invest money into a project and make $242,000 after 2 years at a 10% rate of return. How much money should the Project Manager invest?

A. $50,000
B. $70,000
C. $100,000
D. $200,000

2. A Project Manager wants to invest money into a project and make $330,750 after 2 years at a 5% rate of return. How much money should he/she invest?

   A. $200,000
   B. $300,000
   C. $303,000
   D. $330,000

3. A Project Manager wants to invest money into a project. What is the future value of the project after 2 years, if the present value is $200,000 and the rate of return is 7%?

   A. $200,000
   B. $200,700
   C. $220,000
   D. $228,980

4. A Project Manager wants to invest money into a project. What is the future value of the project after 2 years if the present value is $250,000 and the rate of return is 5%?

   A. $250,000
   B. $250,600
   C. $250,625
   D. $275,625

5. A Project Manager is trying to decide between Project A, which has a net present value of $85,000, and Project B, which has a net present value of $75,000. All other criteria being equal, which project should he/she choose?

   A. Project A
   B. Project B
   C. Both
   D. Neither

6. A Project Manager is trying to decide between Project A, which has a net present value of $35,000, and Project B, which has a net present value of $65,000. All other criteria being equal, which project should he/she choose?

   A. Project B
   B. Project A
   C. Neither
   D. Both

7. A Project Manager is trying to decide between Project A, which has an internal rate of return of 12%, and Project B, which has an internal rate of return of 8%. All other criteria being equal, which project should he choose?

A. Both
B. Neither
C. Project A
D. Project B

8. A Project Manager is trying to decide between Project A, which has an internal rate of return of 17%, and Project B, which has an internal rate of return of 5%. All other criteria being equal, which project should she choose?

A. Neither
B. Both
C. Project B
D. Project A

9. A Project Manager is trying to decide between Project A, which has a payback period of 5 years, and Project B, which has a payback period of 6 years. All other criteria being equal, which project should he choose?

A. Both
B. Neither
C. Project A
D. Project B

10. A Project Manager is trying to decide between Project A, which has a payback period of 15 years, and Project B, which has a payback period of 10 years. All other criteria being equal, which project should she choose?

A. Project A
B. Both
C. Neither
D. Project B

Exercise Answers

1. D

Recall the formula for present value.

Present value = future value $\div$ (1 + internal rate of return)$^{\text{time period}}$

All you have to do is substitute the values. You get $242,000/(1+0.10)^2$. That gives you **$200,000.**

2. **B**

   Recall the formula for present value.

   Present value = future value ÷ (1 + internal rate of return)$^{\text{time period}}$

   All you have to do is substitute the values. You get $330,750/(1 + 0.05)^2$. That gives you **$300,000.**

3. **D**

   Recall the formula for present value.

   Present value = future value ÷ (1 + internal rate of return)$^{\text{time period}}$

   All you have to do is substitute the values. You get future value = $200,000 \times (1 + 0.07)^2$. That gives you **$228,980.**

4. **D**

   Recall the formula for present value.

   Present value = future value ÷ (1 + internal rate of return)$^{\text{time period}}$

   All you have to do is substitute the values. You get future value = $250,000 \times (1 + 0.05)^2$. That gives you **$275,625.**

5. **A**

   Remember that you have to choose the project with the greater net present value. In this case that is **Project A.**

6. **A**

   Remember that you have to choose the project with the greater net present value. In this case that is **Project B.**

7. **C**

   Remember that you have to choose the project with the higher internal rate of return. In this case that is **Project A.**

8. **D**

   Remember that you have to choose the project with the higher internal rate of return. In this case that is **Project A.**

9. **C**

   Remember that you have to choose the project with the shorter payback period. In this case that is **Project A.**

10. **D**

    Remember that you have to choose the project with the shorter payback period. In this case that is **Project B.**

1.2 ACCOUNTING FUNDAMENTALS FOR PROJECT-SELECTION

Main Concept

Financial ratios are used to illustrate and project the financial health of an organization. However, these ratios can also be applied to individual projects, giving the Project Manager valuable insight into the financial feasibility of a project. Certain basic financial ratios can assist a Project Manager in project selection. The Project Manager does not need an accounting background to understand and use these effectively.

Benefit-Cost Ratio

Benefit-cost ratio compares the revenue (benefit) of a project to its cost. Cost-benefit ratio and benefit-cost ratio refer to the same analysis between cost and revenue. The cost-benefit ratio has cost in the numerator and benefit in the denominator. The benefit-cost ratio has benefit in the numerator and cost in the denominator.

- Cost-benefit ratio = cost ÷ benefit
- Benefit-cost ratio = benefit ÷ cost

For a benefit-cost ratio, if the ratio is higher than 1, the project is a profitable venture. If the ratio is equal to 1, the project will break even. If the ratio is less than 1, the costs are higher than the benefit and the project may not be profitable.

For a cost-benefit ratio, if the ratio is higher than 1, the project is not a profitable venture. If the ratio is equal to 1, the project will break even. If the ratio is less than 1, the project is profitable.

Opportunity Cost

Often a company has to choose between projects for reasons of limited capital, resources, or time. Whenever a company has to choose between competing projects, the company has to forgo the profits of the project that was not chosen. Those unrealized profits are the opportunity cost of that selection decision. The opportunity cost is equal to the net present value of the project not chosen. For example, suppose a company chooses Project B with a NPV of $700,000 over Project A with a NPV of $500,000. The opportunity cost of choosing Project B over Project A is $500,000.

Sunk Cost

Sunk costs are the costs that cannot be recovered from a project. While all projects are started in an earnest effort to earn a profit, some projects inevitably veer off course despite a heavy influx of investment and commitment. At some point a management decision must be made regarding the financial feasibility of the project; using the cost-benefit ratio to make that judgment. If midway through the project, the cost outweighs the benefit, it is advisable to write off the amount spent thus far. This investment that cannot be recovered or is not profitable is the sunk cost. Accounting standards dictate that sunk costs should not drive a decision that the project should continue. For example, assume a company has spent $900,000 and still cannot complete a project originally estimated at only $500,000. The project has definitely cost a lot more than originally estimated, but at this point a decision needs to be made if the project is to continue or if it should be scrapped. The sunk cost in this case is the $900,000 that has already been spent and should not be considered while evaluating the project.

Depreciation

Depreciation is an accounting practice that allows a business to spread the cost of an asset over the asset's life span. There are two types of depreciation:

STRAIGHT-LINE DEPRECIATION

In the straight-line type of depreciation, it is assumed that the asset provides the same value in every year of its life span and hence the asset is devalued by the same percentage every year. The calculation is as follows:

Straight-line depreciation = cost of the asset ÷ life of the asset

Assume you have an asset that was originally worth $100,000 and that has an estimated life of 5 years. Straight-line depreciation would be calculated as follows:

	Depreciation Amount	Depreciation Calculation (Cost of the Asset/Life of the Asset)
Year 1	$20,000	100,000 ÷ 5
Year 2	$20,000	100,000 ÷ 5
Year 3	$20,000	100,000 ÷ 5
Year 4	$20,000	100,000 ÷ 5
Year 5	$20,000	100,000 ÷ 5

ACCELERATED DEPRECIATION

With accelerated depreciation, the asset is not devalued by the same percent or amount for every year of its life. Instead, depreciation is higher in the initial years and decreases as time goes on. Accelerated depreciation offers a company several advantages on the accounting front as well as on the tax front. Also, writing off higher amounts in the early years of the asset's life is a more accurate depiction of the asset use. The two popular types of accelerated depreciation are as follows:

- **Declining Balance Method**   Under this method, the depreciation amount is calculated not on the asset's original value but on its current value. To use the declining balance method, use the straight-line method for the first year. For every subsequent year, total up all the depreciation applied until that year and subtract that total from the original asset value. The remainder is considered to be the current asset value. Then divide that amount by the number of years in the asset's estimated total life.

  Current asset value = original asset value – previous depreciations

  Depreciation amt = current asset value ÷ total estimated life

  If an asset was originally worth $100,000 and had an estimated life of 5 years, the depreciation calculated using the declining balance method would be as follows:

	Depreciation Amount	**Depreciation Calculation**
Year 1	$20,000	(100,000 ÷ 5)
Year 2	$16,000	(100,000 – 20,000) ÷ 5
Year 3	$12,800	(100,000 – 36,000) ÷ 5
Year 4	$10,240	(100,000 – 48,800) ÷ 5
Year 5	$8,192	(100,000 – 59,040) ÷ 5

- **Sum of the years method**   In this depreciation method, the life of the asset is of primary importance. First add the numbers from 1 through the estimated number of years in the life of the asset. We'll call this the "sum of the years." To compute the depreciation for any given year, divide the number of years remaining in the estimated life of the asset by the sum of the years. Then multiply the result by the original asset value.

  Depreciation amt = asset value × (remaining estimated life ÷ sum of years)

Once again, consider an asset originally worth $100,000 with an estimated life of 5 years. Using the sum of the years method, depreciation will be calculated as follows:

	Depreciation Amount	Depreciation Calculation
Year 1	$33,333.33	$(100,000) \times (5 \div 15)$
Year 2	$26,666.67	$(100,000) \times (4 \div 15)$
Year 3	$20,000	$(100,000) \times (3 \div 15)$
Year 4	$13,333.33	$(100,000) \times (2 \div 15)$
Year 5	$6,666.67	$(100,000) \times (1 \div 15)$

Return on Sales

Return on sales (ROS) is a measure of a firm's operational efficiency. The firm's net income to sales indicates profit per dollar of sales. Expressed as a percentage, the ROS measures the efficiency of the dollar amount received in sales. It provides valuable insight regarding potential price changes and their impact on profits. This data can be used for competitor price analysis and can be used either to decrease production costs or to increase the price. An increasing ROS over time indicates an increase in efficiency, while a decreasing ROS indicates a decrease in efficiency.

Expenses incurred by a company can be classified as direct, indirect, fixed, and variable costs. The ROS only indicates the need for efficiency, but does not indicate how the efficiency can be achieved.

For example, if the total sale of widgets is $100,000, and the net income before tax is $75,000, the ROS would be 75%.

Return on sales = (net income before or after taxes ÷ total sales) × 100

Return on Assets

Return on assets is a measure of a firm's ability to use its assets efficiently to generate revenue. This ratio is calculated yearly in most companies after they have created their balance sheet. ROA is indicated as a percentage and is calculated as follows:

Return on assets = (net income before or after taxes ÷ total asset value) × 100

ROA is used to gauge the asset intensity of the project or the extent to which the assets contribute to the profits. For instance, a company with a 5% ROA will earn $0.05 for each $1 in assets. ROA and asset intensity have an inverse relation, that is the higher the ROA, the lower the asset intensity. In this case, fewer assets contribute to higher profits.

An example of a company having low ROA would be most manufacturing units that require expensive machinery to build flagship products. These manufacturing units are highly asset intensive, but exhibit lower ROA.

An example of a high ROA could be projects undertaken by strategy and management consulting firms. These firms are not heavily invested in assets and hence exhibit a very high ROA.

Return on Investment

The primary focus of a company is to make sure that all their investments maximize profitability. Return on investment (ROI) is a measure of the profitability of a project from an accounting perspective. ROI is indicated as a percentage and is calculated as follows:

$$\text{Return on investment} = (\text{net income before or after taxes} \div \text{total investment}) \times 100$$

ROI typically includes tangible benefits that translate into a dollar amount. Intangible benefits like brand recognition and goodwill are not accounted for since they have no clearly defined monetary value. The higher the ROI, the more attractive the project is to the company. In most cases companies prefer projects that have a high ROI, but in some cases, a company might select a project that has an ROI less than 1 and that will result in a loss. The reason for doing so might be that the company is a newcomer in the industry and will need to take some losses before it can turn the corner and generate more business and revenue (read break-even point).

As an example, if the net income after taxes of a project is $600,000 and the total investment is $400,000, the return on investment would be

$$(600,000 \div 400,000) \times 100 = 150\%.$$

Working Capital Ratio

Working capital is a measure of a firm's ability to meet its financial obligations toward its liabilities in the short term. It is calculated as follows:

$$\text{Working capital} = \text{current assets} - \text{current liabilities}$$

Working capital can be a positive or a negative number. A positive working capital indicates availability of cash. However, this could be misleading at times. It could be that the company has excessive cash or inventory that accounts for the higher value of assets and hence a higher working capital ratio. In such cases a closer examination of what contributes to the asset value is required. Excessive cash or inventory is not always desirable. A negative working capital indicates excessive liabilities and an inability to meet creditor payments. In case of a negative working capital, a company might have to borrow funds to stay afloat or in some cases even file for bankruptcy.

As an example, if the current assets of a company are $5 million and the current liabilities are $4 million, the working capital would be $1 million. In this case, the current assets are more than current liabilities and the company would seem to be in good financial health.

Break-Even Analysis

Break-even analysis is a tool used to calculate the duration and the revenue that would need to be generated to exceed the costs incurred on a project. Any revenue that is generated after the "break-even point" is the profit. If revenue falls short of the break-even point, the difference constitutes a loss.

Consider a project for building and maintaining a job-application Web site. The project is estimated to cost $300,000 the first year and $100,000 the second year. The yearly maintenance cost has been estimated at $10,000 and revenue expectations are pegged at $50,000 per year. A break-even analysis of this project is as follows:

To figure out when the project will break even, you have to determine when you will recover all the money spent on the project.

$$\text{Static costs of the project} = 300{,}000 + 100{,}000 = \$\,400{,}000$$

Let us assume the number of years taken to break even on this project is x.

so
$$\text{maintenance costs} = 10{,}000x$$

$$\text{Total revenue up to the break-even point} = 50{,}000x$$

You know that at the break-even point, the revenue equals the total costs incurred.

$$\text{Static costs} + \text{maintenance costs} = \text{total revenue up to break-even point}$$

$$400{,}000 + 10{,}000x = 50{,}000x$$
$$40{,}000x = 400{,}000$$
$$x = 10 \text{ years}$$

The conclusion you can reach is that it would take 10 years for the revenue generated by this project to catch up with the costs incurred. After this duration of 10 years the project will be generating profits. This analysis gives management qualitative data that can be used to make informed decisions on the viability and feasibility of undertaking a project.

Accounting Fundamentals for Project-Selection Problems on the PMP Exam

TYPES OF PROBLEMS TO EXPECT

There are six types of problems you can expect in this section:

(1) Determining benefit-cost ratio
(2) Determining opportunity cost
(3) Determining sunk cost
(4) Determining depreciation
(5) Determining break-even analysis
(6) Determining return on sales

DIFFICULTY LEVEL

You do need good memory to remember all the formulas and a thorough understanding of the nuances and differences in the accounting terms discussed to solve these problems. Some of the math using accelerated depreciation methods could become complicated. So work through the sample problems and exercise diligently and it will serve you well for when you are ready to take the exam.

NUMBER OF PROBLEMS TO EXPECT

Expect one problem on accounting ratios on the exam. Most often, this problem will be about the cost-benefit ratio. Other frequent problem topics include opportunity cost and sunk cost.

FORMULAS

- Cost-benefit ratio = cost ÷ benefit
- Benefit-cost ratio = benefit ÷ cost
- Straight-line depreciation = asset value ÷ life of asset
- Declining balance depreciation = current asset value ÷ estimated total life

where current asset talue = original asset value − previous depreciation.

SAMPLE SOLVED PROBLEMS

1. A project has an estimated cost of $200,000 and estimated revenue of $700,000. What is the benefit-cost ratio?

Use the formula:

$$\text{Benefit cost ratio} = \text{benefit} \div \text{cost}$$
$$\text{Benefit cost ratio} = 700{,}000 \div 200{,}000$$
$$\text{Benefit cost ratio} = 7/2$$

2. A new Project A has an estimated NPV of $67,600. However, senior management has decided to pursue a different Project B that has an NPV of $87,600. What is the opportunity cost of selecting Project B?

Management has chosen Project B because it has a higher NPV. The opportunity cost of this choice is equal to the NPV of Project A, which is $67,600.

3. A company purchases capital equipment costing $500,000. The estimated life of the equipment is 5 years. What is the depreciation charged per year if the straight-line depreciation method is used?

Use the formula:

Straight-line depreciation = cost of the asset/life of the asset

Straight-line depreciation = 500,000/5

Straight-line depreciation = $100,000 per year

4. A company purchases capital equipment costing $500,000. The estimated life of that equipment is 5 years. What is the depreciation charged in the third year if the declining balance method of depreciation is used?

Step 1: Find the declining balance method depreciation for the first year.

Declining balance depreciation = current asset value ÷ estimated total life

Declining balance depreciation = 500,000/5

Declining balance depreciation = $100,000

Step 2: Find the declining balance method depreciation for the second year.

Current asset value = original asset value – previous depreciation

Current asset value = 500,000 – 100,000 = 400,000

Declining balance depreciation = current asset value ÷ estimated total life

Declining balance depreciation = 400,000 ÷ 5

Declining balance depreciation = $80,000

Step 3: Find the declining balance method depreciation for the third year.

Current asset value = original asset value – previous depreciation

Current asset value = 500,000 – 180,000 = 320,000

Declining balance depreciation = current asset value ÷ estimated total life

Declining balance depreciation = 320,000 ÷ 5

Declining balance depreciation = $64,000

Hence the depreciation for the third year is $64,000.

5. A company purchases a machine that costs $600,000. The estimated life of that machine is 5 years. What is the depreciation charged in the second year if sum of the years depreciation method is used?

Step 1: Find the sum of the years.

Sum of the years = 5 + 4 + 3 + 2 + 1 = 15

Step 2: Calculate the estimated remaining life of the machine.

After one year, the estimated remaining life of the machine is 5 – 1 = 4 years.

Step 3: Find the second year depreciation.

Depreciation = 4 ÷ 15 × 600,000

Depreciation = $160,000

6. Last year a company's net income before tax was $500,000. Its total sales were $600,000. What was the company's Return on Sales?

Step 1: Find the return on sales.

Return on sales = (net income before tax ÷ total sales) × 100

Return on sales = 500,000 ÷ 600,000 × 100

Return on sales = 83.3%

EXERCISE PROBLEMS

1. A project has an estimated cost of $400,000 and estimated revenue of $600,000. What is the benefit-cost ratio?

    A. 3/2
    B. 4/6
    C. 2/3
    D. 6/4

2. A new Project A has an estimated NPV of $55,600. However, senior management has decided to pursue a different Project B that has an NPV of $67,900. What is the opportunity cost of selecting Project B?

    A. $12,500
    B. $55,600
    C. $67,900
    D. $123,500

3. A construction company purchases a photocopying machine that costs $450,000. The estimated life of that machine is 10 years. What is the depreciation charged per year if the straight-line depreciation method is used?

    A. $4500
    B. $45,000
    C. $90,000
    D. $450,000

4. A widget company purchases a color-printing machine that costs $450,000. The estimated life of that machine is 5 years. What is the depreciation charged in the second year if the declining balance depreciation method is used?

    A. $45,000
    B. $72,000
    C. $90,000
    D. $360,000

5. A widget company purchases a color-printing machine that costs $500,000. The estimated life of that machine is 5 years. What is the depreciation charged in the second year if the sum of the years depreciation method is used?

    A. $130,500.33
    B. $130,900.33
    C. $133,333.33
    D. $135,900.33

6. Last year a car company's net income before tax was $300,000. Its total sales were $400,000. What was the company's return on sales?

    A. 30%
    B. 40%
    C. 75%
    D. 80%

7. A Project Manager is trying to choose between two projects based on their return on investment.

    Project A has projected revenue of $4.5 million and an expected cost of $4 million.

    Project B has projected revenue of $5.5 million and an expected cost of $5 million.

    Which of the following should the Project Manager choose?

    A. Project A with an ROI of 112.50%
    B. Project A with an ROI of 110%
    C. Project B with an ROI of 112.50%
    D. Project B with an ROI of 110%

8. Based on the benefit-cost ratio, which project should a Project Manager choose from the following:

    Project A with a cost of $50 million and expected revenue of $75 million

    Project B with a cost of $100 million and expected revenue of $150 million

    Project C with a cost of $200 million and expected revenue of $300 million

    A. Project A
    B. Project B
    C. All three projects have the same benefit-cost ratio
    D. Project C

9. A certain project costs $500,000 the first year and earns revenue of $100,000 the first year. In its second year, costs are $500,000 and the revenue is $300,000. In each subsequent year, costs are $0 and revenue is $300,000. When will the project break even?

   A. Year 2
   B. Year 4
   C. Year 5
   D. Year 7

10. A certain project requires three software developers at $50 per hour per person for 400 hours, two quality assurance workers at $40 an hour per person for 300 hours, and a Project Manager at $60 an hour for 500 hours. What is the present value of the project?

   A. $24,000
   B. $30,000
   C. $60,000
   D. $114,000

Exercise Answers

1. **A**

   Benefit-cost ratio = benefit ÷ cost. If you plug in the numbers, you get 600,000 ÷ 400,000, and that is **3/2.**

2. **B**

   In this type of model, opportunity cost is equal to the NPV of the project that was not chosen. That amount is **$55,600.**

3. **B**

   Use this formula: straight-line depreciation = asset value/life of the asset

   Here the asset value is 450,000 and the life of the asset is 10 years. The math is simple, and the depreciation per year is **$45,000.**

4. **B**

   Using the declining-balance method, the depreciation in year 1 is $450,000 ÷ 5 = $90,000. The current value of the machine in year 2 is thus $450,000 − $90,000 = $360,000. To find the depreciation for year 2, use this formula: Depreciation current asset value ÷ estimated total life. Plugging in numbers, you get $360,000 ÷ 5 = **$72,000**, which is the depreciation for year 2.

5. C

The sum of the years in this case is 15. So at the start of year 2 there are 4 years left in the life of the asset. $4 \div 15 \times 500,000 =$ **\$133,333.33.**

6. C

The return on sales is net income before or after taxes / total sales. So here it is $300,000 \div 400,000 \times 100 =$ **75%.**

7. A

ROI = (net income before or after taxes ÷ total investment) × 100

Project A has a ROI of $4,500,000 \div 4,000,000 \times 100 = 112.50\%$
Project B has a ROI of $5,500,000 \div 5,000,000 \times 100 =$ **110%**

8. C

Benefit-cost ratio = benefit ÷ cost

Project A has a benefit-cost ratio of $75 \div 50 =$ **1.5**
Project B has a benefit-cost ratio of $150 \div 100 =$ **1.5**
Project C has a benefit-cost ratio of $300 \div 200 =$ **1.5**

So they all have the same benefit-cost ratio. In this case management needs to use other criteria to choose a project from among these three.

9. B

The total cost incurred in the first year is \$500,000 and the revenue is \$100,000. The total cost in year 1 is less than the revenue earned.

By the end of second year, the total cost incurred so far is \$1,000,000 and the total revenue is \$400,000.

By the end of year three, the total cost incurred remains the same \$1,000,000, and the total revenue is now \$700,000. Since the cost and revenue are not yet equal, you have not reached the break-even point.

By the end of year **4**, the total cost incurred remains the same \$1,000,000, and the total revenue is \$1,000,000. Now the cost and revenue are equal, so you have reached the break-even point.

10. D

The cost of each developer is $400 \times 50 = \$20,000$. Because you have three developers, it is \$60,000.

The cost of each quality analyst is $300 \times 40 = 12,000$. Because you have two analysts, the cost is \$24,000.

The cost of the Project Manager is $500 \times 60 = \$30,000$.

Hence the total cost of all the team members is $60,000 + 24,000 + 30,000 =$ **\$114,000.**

1.3 MAKE-OR-BUY DECISION

Main Concept

Once a project is selected, it is time to get a preliminary financial estimate. Make-or-buy analysis simply involves comparing the costs of building the solution in-house to outsourcing it to an outside vendor. A Project Manager has to sum up all the costs involved in creating the solution in-house (if this is an option) and all the costs quoted by the vendor. The success of a make-or-buy decision depends on the ability of the Project Manager to accurately identify and compute all the costs that will be incurred.

"Lease-or-buy" analysis is similar except that only tangible assets can be leased or bought. There are many reasons a company might decide to lease. For example, a company might require a car only for a short time, and in that case it is cheaper to lease than to buy.

Make-or-buy analysis or lease-or-buy analysis is a very important part of a Project Manager's quest for a solution that is cost-effective to the company. Not all projects have an option that involves a make-or-buy decision. Some projects are assigned to the Project Manager after the decision has already been made, in which case the Project Manager can repeat the analysis to ensure that the right decision has been made. If there is a discrepancy, the Project Manager needs to make sure that senior management is aware of the repercussions of their decision.

NOTE *In most projects we find that at some length of time the cost to make equals the cost to buy. This event in time is the break-even point, denoted by z in the two graphs given below (Fig. 1.1).*

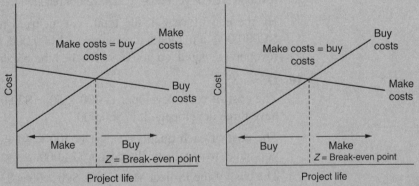

Figure 1.1 (a) Before break-even, the make decision is better; after break-even the buy decision is better; (b) Before break-even, the buy decision is better; after break-even the make decision is better.

As indicated in the graphs (Fig. 1.1*a* and *b*) *z* is used to denote the break-even point. Before and after this particular instant the decision costs differ. If the cost curves for the make and buy costs do not intersect at all, then the decision to choose one over the other becomes a trivial calculation.

An analytical approach can be used to determine the exact time at which these costs are equal.

Let us consider a practical case of analyzing the make-or-buy decision for a software development project. The in-house development costs have been estimated at $1,000,000, with a yearly maintenance cost of $50,000. A third-party vendor has quoted an upfront fee of $500,000 and a yearly maintenance fee of $100,000 for this project.

Make-or-Buy Analysis

The break-even point (in years) is denoted by z, at which the cost of making equals the cost of buying.

$$\text{Cost of making} = 1{,}000{,}000 + 50{,}000 \times z \qquad (1.1)$$

$$\text{Cost of buying} = 500{,}000 + 100{,}000 \times z \qquad (1.2)$$

$$\text{Cost of making} = \text{cost of buying} \quad \text{(at break-even point} = z \text{ years)} \quad (1.3)$$

$$1{,}000{,}000 + 50{,}000z = 500{,}000 + 100{,}000z$$

$$z = 10 \text{ years}$$

This indicates that in exactly 10 years the cost of completing the project in-house (making) equals the cost of buying the software from the vendor.

For any duration less than 10 years, the cost of making is clearly greater than the cost of buying [substitute any value less than 10 for z in Eqs. (1.1) and (1.2)]. In other words, if the life of the project is expected to be less than 10 years it is more economical to buy from a vendor rather than to make in-house.

On the other hand for any duration greater than 10 years, the cost of buying is clearly greater than the cost of making [substitute any value greater than 10 for z in Eqs. (1.1) and (1.2)]. This indicates that it would be a better decision to build or make in-house if you expect the life of the project to exceed 10 years.

Make-or-Buy Decision Problems on the PMP Exam

TYPES OF PROBLEMS TO EXPECT

There are only two types of problems you can expect on this topic:

(1) Computing when a make decision is better than a buy decision and vice versa
(2) Computing when a lease decision is better than a buy decision and vice versa

DIFFICULTY LEVEL

You do not need to have a strong math background to solve these problems. Once you have completed this exercise, you should be ready to face the exam questions.

NUMBER OF PROBLEMS TO EXPECT

Expect at least one problem on make-or-buy analysis on the exam.

FORMULAS

$$\text{Upfront make cost} + (\text{maintenance make cost} \times z)$$
$$= \text{upfront buy cost} + (\text{maintenance buy cost} \times z)$$

where z is the break-even point in time at which the cost of making equals the cost of buying.

INSIDER TIPS

When solving for the number of months or years in a make-or-buy problem, always check your math. Substitute your final answer for z in your original equation. Are the two sides of the equation still equal?

SAMPLE SOLVED PROBLEMS

1. A Project Manager is faced with a make-or-buy decision. A certain software module would cost $48,000 to develop in-house and $3,000 a month to maintain. An outside vendor has offered to provide the module at a charge of $25 per user, and the company

expects around 1,000 users each month. In addition, the outside vendor will charge a maintenance fee of $2,000 per month. When will a buy decision be more advantageous than a make decision?

Step 1: Find the total cost of a make decision.

The development cost will be a one-time cost, but the maintenance cost will be incurred every month, so the total make decision cost is

$$\$48{,}000 + \$3{,}000 \text{ per month of maintenance}$$

Step 2: Find the total cost of a buy decision.

Assume that the number of months that the module will be in service is x

Because the vendor is charging $25 per user and there are 1,000 users per month, this portion of the total expense will be

$$25 \times 1{,}000x = 25{,}000x$$

The $2000 maintenance cost would be incurred every month, so it totals

$$2{,}000x$$

Add the total user charge to the total maintenance fee to find the total cost of the buy decision:

$$25{,}000x + 2{,}000x = 27{,}000x$$

Step 3: Find the point in time at which the cost of a buy decision would equal the cost of a make decision.

Equating the cost of making for z months to the cost of buying for z months (where z is the break-even point), you get

$$48{,}000 + 3{,}000z = 27{,}000z$$

$$27{,}000z - 3{,}000z = 48{,}000$$

$$24{,}000z = 48{,}000$$

$z = 2$, so in this case, z months is 2 months.

At exactly 2 months of maintenance, the cost of making equals the cost of buying.

If the software module is expected to be in service for less than 2 months (let's say 1 month), the cost of buying is $27,000 [Step 2: 27,000 × 1]. The cost of making the module is $51,000 [Step 1: 48,000 + (3,000 × 1)]. Clearly the decision then should be to use the vendor to build the module.

On the other hand, if the module will be in service for more than 2 months (let's say 3 months), the cost of making the module is $57,000 [Step 1: 48,000 + (3,000 × 3)], which is less than the cost of buying it from the vendor for $81,000 [Step 2: 27,000 × 3]. Hence the decision in this case should be to build the module in-house.

EXERCISE PROBLEMS

1. A Project Manager is faced with a make-or-buy decision. The in-house solution costs $23,000 and $2,500 per month to fix defects. A vendor has quoted $15,000 as the upfront charge for purchasing the software and an additional $2,700 as the maintenance fee per month. What is the minimum usage period in months required for the company to justify making the software in-house?

   A. 50
   B. 40
   C. 60
   D. 70

2. A car costs $5,000 cash down to buy and $300 per month in loan payments. The same car can be leased for $1,000 cash and $400 a month. What is the maximum number of months you can lease the car before leasing becomes less economical than buying?

   A. 60
   B. 40
   C. 50
   D. 70

3. A Project Manager needs to develop a web site that allows users to submit expenses. There are two vendors competing for the project. Vendor A proposes an upfront fee of $20,000 and a maintenance fee of $4,000 per month. Vendor B proposes an upfront fee of $10,000 and a maintenance fee of $5,000 per month. Which vendor will be a better choice and after how many months of maintenance?

   A. Vendor A after 10 months of maintenance
   B. Vendor B after 10 months of maintenance
   C. Vendor A after 5 months of maintenance
   D. Vendor B after 5 months of maintenance

4. Given the information below, after how many months will a make decision be better than a buy decision?

   A. 15
   B. 20
   C. 25
   D. 40

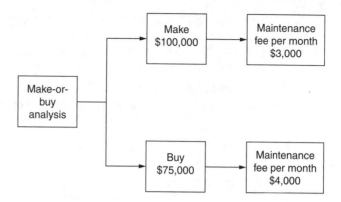

5. Given the information below, after how many months will a buy decision be better than a make decision?

    A. 7
    B. 8
    C. 9
    D. 10

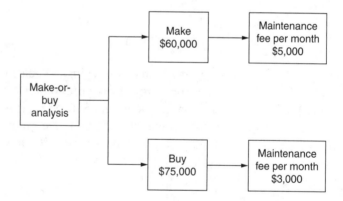

6. A company is faced with a make-or-buy decision for a software component. The in-house solution costs $14,000 and $2,000 per month to fix defects. A vendor has quoted $5,000 as the upfront cost for purchasing the software and an additional $5,000 as a maintenance fee per month. For how many months should the company plan to use the software to support a make decision?

    A. 3
    B. 4
    C. 5
    D. 6

7. A Project Manager needs to develop a software module to interface with a legacy system. Hiring a full-time employee to do the job requires $45,000 as a hiring cost and a monthly salary of $5,000. Hiring a consultant would require a $10,000 initial fee and $6,000 per month. After how many months of work will it be more profitable to hire an employee over a consultant?

A. 10
B. 25
C. 35
D. 45

8. A Project Manager is trying to decide between Software A and Software B for a project assigned to her. Software A costs $75,000 and there is a $5,000 maintenance fee per month. Software B costs $25,000 and there is a $15,000 maintenance fee per month. What is the minimum number of months of usage for which Software A would be the cheaper alternative?

A. 5
B. 10
C. 15
D. 25

9. A Project Manager is trying to decide between Vendor A and Vendor B for a project assigned to him. Vendor A charges $25,000 as an initial payment and $10,000 per month as a maintenance fee. Vendor B charges $55,000 as an initial payment and $5,000 per month as a maintenance fee. After how many months would Vendor B's plan be cheaper?

A. 5
B. 6
C. 7
D. 8

10. Given the information below, after how many months will Vendor A be a better (cheaper) choice than Vendor B?

A. 4
B. 5
C. 6
D. 7

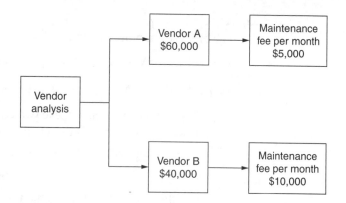

Exercise Answers

1. **B**

   The break-even point is denoted by z months of maintenance.

   Step 1: Cost of make decision $= 23,000 + 2,500z$

   Step 2: Cost of buy decision $= 15,000 + 2,700z$

   Step 3: Now you know that at the break-even point (z months), the decision costs are equal.

   Hence, $\qquad 23,000 + 2500z = 15,000 + 2,700z$

   $$z = \textbf{40 months}$$

   Step 4: Substituting any value greater than $z = 40$ months in the decision-cost equations of Step 1 and Step 2 clearly indicates that the cost of an in-house solution is less than the buy cost when the solution is expected to be in service for a minimum of 40 months.

2. **B**

   The break-even point is denoted by z months of car usage.

   Step 1: Cost of buying car $= 5,000 + 300z$

   Step 2: Cost of leasing car $= 1,000 + 400z$

   Step 3: Now you know that at the break-even point (z months), the cost of leasing and buying are equal.

   Hence, $\qquad 5,000 + 300z = 1,000 + 400z$

   $$z = \textbf{40 months}$$

   Step 4: Substituting any value less than $z = 40$ months in the decision-cost equations of Step 1 and Step 2 clearly indicates that the cost of leasing the car is less than the cost of buying it as long as you use the use the car for a maximum of 40 months.

3. **A**

The break-even point is denoted by z months of maintenance.

Step 1: Cost of Vendor A $= 20{,}000 + 4{,}000z$

Step 2: Cost of Vendor B $= 10{,}000 + 5{,}000z$

Step 3: Now you can equate these two costs to find the break-even point at which the costs of both vendors are equal.

$$20{,}000 + 4{,}000z = 10{,}000 + 5{,}000z$$

$$z = \textbf{10 months}$$

So at exactly 10 months of maintenance, the costs of both vendors are equal.

Step 4: Substituting any value less than $z = 10$ months in the vendor-cost equations of Step 1 and Step 2 clearly indicates that the cost of Vendor A is more than the cost of Vendor B and hence Vendor B would be the best choice if the estimated maintenance period is a maximum of 10 months.

What this also means is that after the end of 10 months, Vendor A's cost is less than that of Vendor B (substitute any value greater than $z = 10$ months in the vendor-cost equations of Step 1 and Step 2). So if the estimated maintenance period is more than 10 months, Vendor A is the best choice.

4. **C**

The key here is to find the break-even point, denoted by z months of maintenance.

Step 1: Cost of making $= 100{,}000 + 3{,}000z$

Step 2: Cost of buying $= 75{,}000 + 4{,}000z$

Step 3: Now you can equate these two costs to find the break-even point at which the costs are equal.

$$100{,}000 + 3{,}000z = 75{,}000 + 4000z$$

$$z = \textbf{25 months}$$

The break-even point is 25 months.

Step 4: Substituting any value greater than $z = 25$ months in the decision-cost equations of Step 1 and Step 2 clearly indicates that the cost of making becomes less than the cost of buying as soon as the break-even point of 25 months is reached.

5. B

The key here is to find the break-even point, denoted by z months of maintenance.

Step 1: Cost of making $= 60,000 + 5,000z$

Step 2: Cost of buying $= 75,000 + 3,000z$

Step 3: Now you can equate these two costs to find the break-even point at which the costs are equal.

$$60,000 + 5,000z = 75,000 + 3,000z$$

$$z = \textbf{7.5 months}$$

The break-even point is 7.5 months.

Step 4: Substituting any value greater than $z = 7.5$ months in the decision-cost equations of Step 1 and Step 2 clearly indicates that the cost of buying becomes less than the cost of making as soon as the break-even point of 7.5 months is reached. The answer choice that is closest to 7.5 and greater than 7.5 is choice B (8 months).

6. A

The break-even point is denoted by z months of maintenance, used primarily to manage defects.

Step 1: Cost of making $= 14,000 + 2,000z$

Step 2: Cost of buying $= 5,000 + 5,000z$

Step 3: Now you know that at the break-even point (z months), the costs of making and buying are equal.

Hence, $14,000 + 2,000z = 5,000 + 5,000z$

$$z = \textbf{3 months}$$

Step 4: Substituting any value greater than $z = 3$ months in the decision-cost equations of Step 1 and Step 2 clearly indicates that the cost of making is less than the cost of buying if the maintenance period of the software component is a minimum of 3 months. Any period less than 3 months favors a buy decision.

7. C

The break-event point is denoted by z months.

Step 1: Cost of hiring employee $= 45,000 + 5,000z$

Step 2: Cost of hiring consultant $= 10,000 + 6,000z$

Step 3: Now you know that at the break-even point (z months), the cost of hiring an employee is the same as the cost of hiring a consultant.

Hence, $45,000 + 5000z = 10,000 + 6,000z$

$$z = \textbf{35 months}$$

Step 4: Substituting any value greater than $z = 35$ months in the decision-cost equations of Step 1 and Step 2 clearly indicates that the cost of hiring an employee is less than the cost of hiring a consultant. So it is definitely preferable to hire an employee if the project is expected to last at least **35 months.**

Using any value less than 35 in the cost equations indicates that it is preferable to use a consultant if the project is expected to last less than 35 months.

8. A

The break-event point is denoted by z months of software maintenance.

Step 1: Cost of choosing Software A $= 75,000 + 5,000z$

Step 2: Cost of choosing Software B $= 25,000 + 15,000z$

Step 3: Now you know that at the break-even point (z months), the costs of choosing Software A and Software B are equal

Hence, $75,000 + 5,000z = 25,000 + 15,000z$

$$z = \textbf{5 months}$$

Break-even point is 5 months of software maintenance.

Step 4: Substituting any value greater than $z = 5$ months in the decision-cost equations of Step 1 and Step 2 clearly indicates that the cost of choosing Software A is less than the cost of choosing Software B. So prior to the break-even point (i.e., less than 5 months), it is cheaper to use Software B.

The correct answer is choice A, because it is cheaper to choose Software A only if the maintenance period is estimated to be at least 5 months.

9. B

The break-event point is denoted by z months of project maintenance.

Step 1: Cost of choosing Vendor A $= 25,000 + 10,000z$

Step 2: Cost of choosing Vendor B = $55,000 + 5,000z$

Step 3: Now you know that at the break-even point (z months), the costs of choosing Vendor A and Vendor B are equal.

Hence, $\qquad 25,000 + 10,000z = 55,000 + 5,000z$

$$z = \textbf{6 months}$$

The break-even point is 6 months of project maintenance.

Step 4: Substituting any value greater than $z = 6$ months in the decision-cost equations of Step 1 and Step 2 clearly indicates that the cost of choosing Vendor B is less than the cost of choosing Vendor A. Prior to the break-even point (i.e., less than 6 months), it is cheaper to choose Vendor A.

The correct answer is choice B because it is cheaper to choose Vendor B only when the project maintenance period is estimated to be at least 6 months.

10. A

The key here is to find the break-even point, denoted by z months at which costs of both vendors are equal.

Step 1: Cost of Vendor A = $60,000 + 5000z$

Step 2: Cost of Vendor B = $40,000 + 10,000z$

Step 3: Now you can equate these two costs to find the break-even point at which the costs are equal.

$$60,000 + 5,000z = 40,000 + 10,000z$$

$$z = \textbf{4 months}$$

The break-even point is 4 months.

Step 4: Substituting any value greater than $z = 4$ months in the vendor-cost equations of Step 1 and Step 2 clearly indicates that Vendor A's cost beats (is less than) Vendor B's cost. This means that Vendor A is a better (cheaper) choice after a minimum of 4 months.

CHAPTER 2

Planning

This chapter covers	1. Probability fundamentals
	2. Planning tools and techniques
	• Expected monetary value
	• Decision tree analysis
	• Project cost estimation
	• Critical path analysis and float
	• Crashing
	• PERT, standard deviation and variance
	• Contract cost estimation
	• Point of total assumption
	• Communication channels

INTRODUCTION

Planning is the most crucial phase of a project, and it is here that an effective Project Manager can have a profound impact on the outcome. The planning process results in the creation of a realistic and formal Project Management Plan. This plan has several components that describe how the project's scope, schedule, cost, quality, and resources will be planned and controlled. Planning also helps flesh out the scope of the project. Planning is an iterative process, and the Project Management Plan is progressively elaborated as the project evolves and data integrity improves. The iterative process also accounts for and adapts to unforeseen events or results and scope modifications.

A Project Manager has to plan proactively to meet the project budget and schedule. Even unforeseen events have to be accounted for in the budget and schedule by providing contingency reserves.

In the sections that follow, we will introduce some key mathematical concepts that are relevant to the planning process and also walk through techniques for schedule network analysis, schedule compression, project risk analysis, contract cost administration, and the like. Each of these concepts are defined here but will be described in detail later in this chapter.

- **Probability**   Probability is an important concept to consider when analyzing risks. A Project Manager has to thoroughly analyze the possible risks that will hinder or augment the team's ability to meet the scheduled due date and the budget.
- **Expected Monetary Value**   Expected Monetary Value (EMV) is driven by probability and assigns an expected value to each risk.
- **Decision Tree**   Both probability and EMV are used to illustrate the likelihood and impact of all possible decisions. These decisions are represented graphically in a tree-like illustration (a "decision tree") for easy visual assessment.
- **Project Cost Estimate**   An important aspect of planning is estimating the cost of the project. Risk probability, EMV, and the decision tree are used to determine the amount of contingency and management reserve that needs to be added to the cost estimate.
- **Critical Path**   Critical path analysis is used to determine the key project tasks that need to be accomplished and the shortest duration required to complete them by multitasking. Float extends the critical path analysis concept by gauging the amount of time an activity can be delayed without extending the project finish date.
- **Crashing**   Crashing involves weighing options to get optimal schedule compression with the option that costs the least.
- **PERT**   PERT is a widely used schedule-estimation tool. PERT arrives at a more realistic estimate by assigning a weight to optimistic, pessimistic, and most likely estimates for each task. The probability of meeting the PERT-estimated schedule increases with each standard deviation.
- **Cost of Contracts**   Cost of contracts plays a very vital role when outsourcing work on a project to a third-party vendor.
- **Point of Total Assumption**   Point of total assumption (PTA) defines the breakpoint beyond which the seller assumes responsibility for all project costs. It is helpful to know how to calculate PTA for fixed price or cost reimbursable contracts.
- **Communication Channels**   Considering that 90% of a Project Manager's time is spent communicating, it is vital to know how communication channels contribute to a project's success.

2.1 PROBABILITY FUNDAMENTALS

Main Concept

In discussing probability, mathematicians and others use the following conventions:

- The probability of an event can be denoted as p(event), probability (event), or P(event). In this book, we will use p(event) as the standard notation of probability.
- Venn diagrams are used when necessary to illustrate probability.

TIP: *Need help with understanding Venn diagrams? Refer to Appendix A for a brief introduction of Venn diagrams.*

Have you been using any excuse you could think of to avoid math classes since high school? If so, now is the time to set aside your math phobia and read this chapter with an open mind. What is probability and why is it so pertinent to project management? Because you as a Project Manager will ultimately be responsible for the success of a project, you will need to analyze what risks might adversely affect the project, and what factors might boost its outcome. Once you identify a potential risk event, you will need to conduct a thorough quantitative analysis to calculate the odds of that event occurring and its impact on the project. This is where probability comes into play—the quantitative analysis of risks, threats, and opportunities. This analysis will help you to focus on risks that merit immediate attention and to prioritize them. A Project Manager cannot deal with or preempt all risks, but risk analysis can help in making informed project decisions. Risks can be good or bad, opportunities or threats. Knowing the odds of an opportunity or threat occurring is critical to project success. Let us review some fundamental concepts of probability.

Probability of an Event Occurring

The probability of an event occurring is mathematically defined as the ratio of the number of favorable outcomes to the total number of possible outcomes. Probability will always be a number greater than or equal to 0 and less than or equal to 1.

For example, suppose you roll a die with six faces marked with the numbers 1 to 6. The die could land with any one of the six faces on top, so the total number of possible outcomes is 6. What is the probability of rolling the number 4? Only one face of the die has the number 4 on it, so the number of possible "favorable" outcomes (i.e., rolling a 4) is 1. The probability of rolling a 4 is the ratio of the number of possible favorable outcomes (1) to the total number of possible outcomes (6), or 1/6 (See Fig. 2.1).

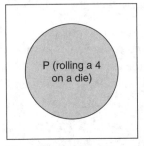

Figure 2.1 Probability of rolling the number 4 on a die.

Probability of an Event Not Occurring

The probability of an event *not* occurring is mathematically defined as the ratio of the number of unfavorable outcomes to the total number of outcomes. Suppose once again that you are rolling a die with six faces. What is the probability of *not* rolling a 4? There are still a total of 6 possible outcomes of rolling the die. Five of the faces show a number other than 4, so the number of possible unfavorable outcomes (i.e., rolling a number other than 4) is 5. Thus the probability of not rolling a 4 is the ratio of the number of unfavorable outcomes (5) to the total number of possible outcomes (6), or 5/6. This ratio can also be calculated as 1 − (probability of rolling a 4) or 1 − 1/6 = 5/6 (See Fig. 2.2).

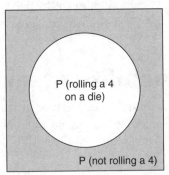

Figure 2.2 Probability of *not* rolling the number 4 on a die.

Mutually Exclusive Events

Mutually exclusive events are events that cannot occur simultaneously. For example, suppose that you are now rolling a pair of dice. For a single roll of the pair of dice, you want to determine the probability that the sum of the numbers on the faces of the dice will be either 3 or 5. These two outcomes, or events, are mutually exclusive ("disjoint") because they cannot occur at the same time (i.e., on a single roll of the two dice). The sum of the numbers can either be 3 or 5 or some other number, but the numbers can never add up to 3 and 5 on the same roll (See Fig. 2.3). For mutually exclusive events A and B,

$$p \, (A \text{ or } B) = p \, (A) + p \, (B)$$

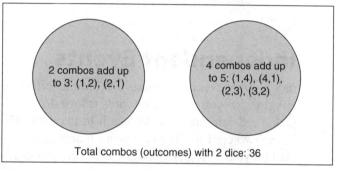

Figure 2.3 Probability of rolling two numbers that add up to either 3 or 5 on a single roll of two dice.

Events That Are Not Mutually Exclusive

Events that are not mutually exclusive are events that can occur simultaneously and have some outcomes in common. Suppose once again that you are rolling a pair of dice. For a single roll of the pair of dice, you want to determine the probability of rolling two numbers that add up to either 6 or an even number (See Fig. 2.4). These two events are not mutually exclusive or disjoint because by definition, any two numbers that add up to 6 also add up to an even number. For events A and B that are not mutually exclusive,

$$p \, (A \text{ or } B) = p \, (A) + p \, (B) - p \, (A \text{ and } B)$$

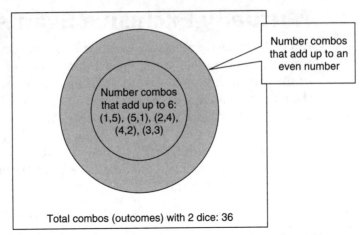

Figure 2.4 Probability of rolling two numbers that add up to either six or an even number.

Independent Events

Independent events are events that have no effect on each other. For example, imagine that you are picking cards from a deck of 52 cards. Suppose you first pick a red card and then replace it in the deck. Then you pick a black card. These two card-picking events are independent of each other. Picking a red card on your first pick does not affect the probability of picking a black card on your second pick because in both cases you are picking from a complete deck of 52 cards. In both cases, you are picking from the exact same number of red and black cards.

Independent events can be joint or disjoint events. So it is not possible to represent them accurately in a Venn diagram. For two independent events A and B,

$$p \text{ (A and B)} = p \text{ (A)} \times p \text{ (B)}$$

Dependent Events

Dependent events are events that can occur in sequence and that influence each other. Consider again the act of picking cards from a deck of 52 cards. Suppose that on your first pick you pick a red card, but you do *not* replace it in the deck. You then make your second pick from the remaining 51 cards. In this case, the probability of picking a black card on your second pick is affected by your first pick; the second event is dependent on the first. If your first pick is red, the probability of picking a black card on the second pick is greater than the probability of picking a red card because there are now 25 red cards and 26 black ones remaining in the deck. However, if your first pick is black, the probability of picking a black card on your second pick is less than the probability of picking a red card because there are now 26 red cards and 25 black cards remaining in the deck. Dependent events are outside the scope of the PMP exam, but this is an important concept to understand.

Probability Fundamentals Problems on the PMP Exam

TYPES OF PROBLEMS TO EXPECT

There are five types of probability problems that you can expect on the PMP exam:

(1) Computing the probability of an event occurring
(2) Computing the probability of an event not occurring
(3) Computing the probability of mutually exclusive events occurring (*either/or*)
(4) Computing the probability of events that are not mutually exclusive occurring (*either/and*)
(5) Computing the probability of independent events occurring (*and*)

DIFFICULTY LEVEL

This topic is one of the most challenging on the PMP exam, especially if you do not have a strong math background or if your math skills have eroded over time. Understand the fundamentals described in this section and work through the exercises in this book and the CD to prepare yourself to handle any problem related to probability.

NUMBER OF PROBLEMS TO EXPECT

Expect at least one problem regarding probability on the exam.

FORMULAS

Let us review the formulas for probability problems:

- **Probability of an event occurring**

p (event A) = number of favorable outcomes/total number of possible outcomes

- **Probability of an event *not* occurring**

    p (event A) not occurring = 1 − p (event A)

- **Probability of two mutually exclusive events occurring**

    p (A or B) = p (event A) + p (event B)

- **Probability of two events that are not mutually exclusive occurring**

    p (A or B) = p (event A) + p (event B) − p (A and B)

- **Probability of two independent events occurring**

    p (A and B) = p (event A) × p (event B)

- **Probability of two dependant events occurring (conditional probability)**

    p (A and B) = p (event A) × p(B|A)

INSIDER TIPS

- If probability is expressed as a ratio, it always ranges between 0 and 1. So eliminate any answer choice that is less than 0 or greater than 1. Also, probability can never be equal to 1 if there is more than 1 possible outcome.

- If probability is expressed as a percentage, just multiply the probability ratio that you calculated by 100. The result will be a number between 0 and 100 percent.

- Read the problem twice to determine if the events are mutually exclusive or independent. This will tell you which formula to apply.

- Remember to *multiply* the probability of the events if you see the word "and" or if you see the words "independent events."

- Remember to *add* the probability of the events if you see the word "or" or if you see the words "mutually exclusive events."

SAMPLE SOLVED PROBLEMS

1. On a single roll of a six-sided die, what is the probability of rolling a 5?

Step 1: Find the total possible outcomes when the die is rolled.

That would be 6 because the die has six sides numbered 1, 2, 3, 4, 5 and 6, respectively.

Step 2: Find the total possible outcomes in which a 5 is rolled.

That would be 1 because only one side of the die is labeled with a 5.

Step 3: Compute the probability.

Use the formula:

p (event A) = number of favorable outcomes/total number of possible outcomes

Thus the probability is 1/6 = 0.167.

Step 4: Compute the probability as a percentage (if required).

Multiply the probability by 100.

$0.167 \times 100 = 16.7\%$, which means that there is a 16.7% chance of rolling the number 5 on one roll of a die.

2. On a single roll of a six-sided die, what is the probability of *not* rolling a 6?

Step 1: Find the total possible outcomes when the die is rolled.

That would be 6 because the die has six sides numbered 1, 2, 3, 4, 5, and 6, respectively.

Step 2: Find the total possible outcomes in which a 6 is rolled.

That would be 1 because only one side of the die is labeled with a 6.

Step 3: Compute the probability of the event occurring.

p (event A) = number of favorable outcomes/total number of possible outcomes

Thus the probability of rolling a 6 is 1/6 = 0.167.

Step 4: Compute the probability of the event not occurring.

To calculate the probability of not rolling a 6, use the formula:

p (event A) not occurring = 1 − p (event A)

So the probability of not rolling a 6 is $1 - 0.167 = 0.833$.

Step 5: Compute the probability as a percentage (if required).

Multiply the probability by 100.

$0.833 \times 100 = 83.3\%$, which means that there is an 83.3% chance of not rolling a 6 on one roll of a die.

3. When rolling a pair of dice, what is the probability of rolling two numbers that add up to either 3 or 6?

To find the probability,

Step 1: Find the total number of possible outcomes when a pair of dice is rolled.

There are 36 possible outcomes. They are:

(1,1)	**(2,1)**	(3,1)	(4,1)	(5,1)	(6,1)
(1,2)	(2,2)	(3,2)	(4,2)	(5,2)	(6,2)
(1,3)	(2,3)	(3,3)	(4,3)	(5,3)	(6,3)
(1,4)	(2,4)	(3,4)	(4,4)	(5,4)	(6,4)
(1,5)	(2,5)	(3,5)	(4,5)	(5,5)	(6,5)
(1,6)	(2,6)	(3,6)	(4,6)	(5,6)	(6,6)

Step 2: Find the probability of rolling two numbers that add up to 3.

There are 2 possible outcomes in which the numbers add up to 3. They are (1, 2) and (2, 1), shown in bold in the table above. Call this event A. The probability of event A is

(Number of favorable outcomes)/(total number of possible outcomes) = 2/36.

Step 3: Find the probability of rolling two numbers that add up to 6.

There are 5 possible outcomes in which the numbers add up to 6. They are (1, 5), (2, 4), (3,3), (4,2), and (5,1), all of which are underlined in the preceding table above. Call this event B. The probability of event B is

(Number of favorable outcomes)/(total number of possible outcomes) = 5/36.

Step 3: Find the probability of rolling two numbers that add up to either 3 or 6.

Note that event A and event B are mutually exclusive because you can never have a single outcome in which the sum of the numbers rolled is both 3 and 6. So use the formula:

$$p \text{ (A or B)} = p \text{ (event A)} + p \text{ (event B)}$$

Probability of event A or event B occurring = 2/36 + 5/36 = 7/36

Probability of event A or event B occurring $\approx$ 0.195 (The symbol $\approx$ means "is approximately equal to.")

Step 4: Compute the probability as a percentage (if required).

Multiply the probability by 100.

$0.195 \times 100 = 19.5\,\%$; which means that there is about a 19.5% probability that the two numbers rolled will add up to either 3 or 6.

4. A pair of dice is rolled. What is the probability that the sum of the numbers rolled is either an even number or a multiple of 3?

To find the probability,

Step 1: Find the total number of possible outcomes when a pair of dice is rolled.

There are 36 possible outcomes. They are:

(1,1)	(2,1)	**(3,1)**	(4,1)	**(5,1)**	(6,1)
(1,2)	**(2,2)**	(3,2)	**(4,2)**	(5,2)	**(6,2)**
(1,3)	(2,3)	**(3,3)**	(4,3)	**(5,3)**	(6,3)
(1,4)	(2,4)	(3,4)	**(4,4)**	(5,4)	**(6,4)**
(1,5)	(2,5)	**(3,5)**	(4,5)	**(5,5)**	(6,5)
(1,6)	**(2,6)**	(3,6)	**(4,6)**	(5,6)	(6,6)

Step 2: Find the probability of rolling two numbers that add up to a multiple of 3.

There are 12 possible outcomes in which the numbers rolled add up to a multiple of 3. They are (1,2), (1,5), (2,1), (2,4), (3,3), (3,6), (4,2), (4,5), (5,1), (5,4), (6,3), and (6,6), all underlined in the preceding table above. Call this event A. The probability of event A is

(Number of favorable outcomes)/(total number of possible outcomes) = 12/36.

Step 3: Find the probability of rolling two numbers that add up to an even number.

There are 18 possible outcomes in which the numbers rolled add up to an even number. They are (1,1), (1,3), (1,5), (2,2), (2,4), (2,6), (3,1), (3,3), (3,5), (4,2), (4,4), (4,6), (5,1), (5,3), (5,5), (6,2), (6,4), and (6,6), shown in boldface in the table above. Call this event B. The probability of event B is

(Number of favorable outcomes)/(total number of possible outcomes) = 18/36.

Step 4: Find the probability of rolling two numbers that add up to a multiple of 3 which is *also* an even number.

Note that event A and event B are not mutually exclusive because some number pairs add up to a number that is both a multiple of 3 and an even number. There are 6 possible outcomes in this category. They are (1,5), (2,4), (3,3), (4,2), (5,1), and (6,6), shown underlined and in boldface in the preceding table. Call this event (A and B). The probability of event (A and B) is

(Number of favorable outcomes)/(total number of possible outcomes) = 6/36.

Step 5: Find the probability of rolling two numbers that add up to either a multiple of 3 or an even number. Use the formula:

p (A or B) = p (event A) + p (event B) − p (A and B)

p (A or B) = p (12/36) + p (18/36) − p(6/36)

p (A or B) = 30/36 − 6/36 = 24/36 ≈ 0.67

Step 6: Compute the probability as a percentage (if required).

Multiply the probability by 100.

0. 67 × 100 = 67%; which means that there is about a 67% chance that the sum of the numbers rolled will be either an even number or a multiple of 3.

5. What is the probability of picking a red card from a deck of 52 cards and then replacing the card and picking a black card from the deck of 52 cards?

To find the probability,

Step 1: Find the probability of picking a red card from a deck of 52 cards.

Call this event A. The number of red cards in a deck of cards is 26 (13 hearts and 13 diamonds). So

p (event A) = number of red cards/total number of cards in the deck

$$= \frac{26}{52} = \frac{1}{2} = 0.5$$

Step 2: Find the probability of picking a black card from a deck of 52 cards.

Call this event B. The number of black cards in a deck of cards is 26 (13 spades and 13 clubs). Note that the first (red) card picked is replaced in the deck, so the deck still totals 52 cards.

So p (event B) = number of black cards/total number of cards in the deck

$$= \frac{26}{52} = \frac{1}{2} = 0.5$$

Step 3: Find the probability of picking a red card from a deck of 52 cards and then a black card from the deck of 52 cards.

These are two independent events because the result of the first pick does not in any way affect the result of the second pick. Use the formula:

$$p \text{ (A and B)} = p \text{ (event A)} \times p \text{ (event B)}$$

$$p \text{ (red and black)} = 0.5 \times 0.5 = 0.25$$

Step 4: Compute the probability as a percentage (if required).

Multiply the probability by 100. $0.25 \times 100 = 25\%$; which means that there is a 25% chance of first picking a red card from a deck, replacing it in the deck, and then picking a black card.

6. What is the probability of picking a red card from a deck of 52 cards and then choosing a black card from the same deck without replacing the red card?

Step 1: Find the probability of picking a red card from a deck of 52 cards.

Call this event A. The number of red cards in a deck of cards is 26 (13 hearts and 13 diamonds). So

p (event A) = number of red cards/total number of cards in the deck
$$= \frac{26}{52} = \frac{1}{2} = 0.5$$

Step 2: Find the probability of picking a black card from a deck of 52 cards after a red card has already been picked.

Call this event B|A. Note that the first (red) card picked has *not been* replaced in the deck, so the deck now totals only 51 cards: 25 red and 26 black. So

p (event B|A) = number of black cards/total number of cards in the deck
$$= 26/51 = 0.509$$

Step 3: Find the probability of first picking a red card from a deck of 52 cards and then picking a black card without replacing the first pick. These are dependent events, so use the formula:

$$p \text{ (A and B)} = p \text{ (event A)} \times p \text{ (B|A)}$$
$$p \text{ (A)} = 26/52 = 0.5$$
$$p \text{ (B|A)} = 26 / 51 = 0.509$$

So, p (red and black) $= 0.5 \times 0.509 \approx 0.2545$.

Step 4: Compute the probability as a percentage (if required).

All you need to do is multiply the probability by 100.

$$0.2545 \times 100 = 25.45\%.$$

---

7. In the following network diagram, tasks A, B, and C have to be completed before Task D can be started. Task A has a 60% chance of being completed on day 10, Task B has a 50% chance of being completed on day 10, and Task C has an 80% chance of being completed on day 10. What is the probability of starting Task D on day 11?

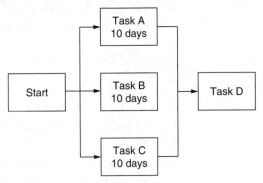

Step 1: Calculate the probability of tasks A, B, and C being completed on day 10.

$$p \text{ (Task A)} \times p \text{ (Task B)} \times p \text{ (Task C)}$$
$$(60/100) \times (50/100) \times (80/100) = 24\%$$

So the probability of starting Task D on Day 11 is 24%.

EXERCISE PROBLEMS

1. The probability of meeting a scheduled deadline is 25%. What is the probability of not meeting the deadline?

   A. 25%
   B. 5%
   C. 75%
   D. 50%

2. In a new project of developing a financial application for a bank, employee attrition and scope change are two independent events. The probability of employee attrition is 20%, and the probability of scope change is 30%. What is the probability of both these events occurring?

   A. 6%
   B. 10%
   C. 50%
   D. 60%

3. The probability of not meeting a project's estimated cost is 30%. What is the probability of completing the project on budget?

   A. 25%
   B. 70%
   C. 100%
   D. 30%

4. Assume that making a certain product and buying it are mutually exclusive events. If the probability of making the product in-house is 25% and the probability of buying it is 15%, what is the probability that either of these events will occur?

   A. 15%
   B. 20%
   C. 25%
   D. 40%

5. As a Project Manager handling a construction project, you begin evaluating the risks. Rain and snow are two independent events that affect the schedule the most. If the risk (probability) of rain is 15% and the risk of snow is 5%, what is the probability that the project will be affected by both rain and snow?

   A. 0.75%
   B. 10%
   C. 20%
   D. 75%

6. Assume that meeting the schedule deadline and staying within the budget are independent events. If the probability of meeting the schedule is 10%, but the probability of staying within budget is

only 5%, what is the probability of completing the project on time and on budget?

A. 50%
B. 15%
C. 5%
D. 0.5%

7. In the preceding network diagram, tasks A and B both have to be completed before Task C can be started. Task A has a 20% chance of being completed on day 10, and Task B has a 25% chance of being completed on day 10. What is the probability of starting Task C on day 11?

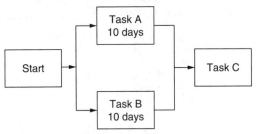

A. 0.05%
B. 0.5%
C. 5%
D. 50%

8. In the network diagram shown, tasks A, B, and C have to be completed before Task D can be started. Task A has a 40% chance of being completed on day 15, Task B has a 50% chance of being completed on day 15, and Task C has a 60% chance of being completed on day 15. What is the probability of starting Task D on day 16?

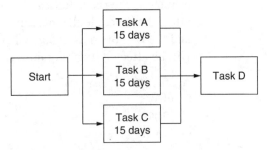

A. 50%
B. 12%
C. 1.2%
D. 0.12%

Exercise Answers

1. **C**

   The probability of an event *not* occurring is mathematically defined as the ratio of the number of unfavorable outcomes to the total number of possible outcomes. In this case you are told that the probability of meeting the scheduled deadline is 25%, or 0.25. So the probability of not meeting the deadline is $1 - 0.25$, which is 0.75 or **75%**.

2. **A**

   Recall that the probability of two independent events occurring is calculated using the formula
   $$p \text{ (A and B)} = p \text{ (event A)} \times p \text{ (event B)}$$
   In this case, $20\% \times 30\% = (20/100) \times (30/100) = 600/10{,}000 = \mathbf{6\%}$.

3. **B**

   In this case you are told that the probability of *not* meeting the cost estimate is 30%. Thus the probability of meeting the estimate and completing the project on budget is $1 - 0.3 = 0.7 = \mathbf{70\%}$.

4. **D**

   The probability of two mutually exclusive events occurring is calculated using the formula
   $$p \text{ (A or B)} = p \text{ (event A)} + p \text{ (event B)}$$
   In this case, $25\% + 15\% = \mathbf{40\%}$.

5. **A**

   Because rain and snow are independent events, at least for the sake of this problem,
   $$p \text{ (A and B)} = p \text{ (event A)} \times p \text{ (event B)}$$
   $p \text{ (A and B)} = 15\% \times 5\% = (15/100) \times (5/100) = 75/10{,}000 = \mathbf{0.75\%}$.

6. **D**

   For the purposes of this problem, meeting the schedule deadline and staying within budget are independent events. Use the formula:
   $$p \text{ (A and B)} = p \text{ (event A)} \times p \text{ (event B)}$$
   $p \text{ (A and B)} = 10\% \times 5\% = (10/100) \times (5/100) = 50/10{,}000 = \mathbf{0.5\%}$.

7. **C**

   The probability of starting Task C on day 11 is $p \text{ (A)} \times p \text{ (B)} = (20/100) \times (25/100) = 500/10{,}000 = \mathbf{5\%}$.

8. **B**

   The probability of starting Task D on day 16 is $p \text{ (A)} \times p \text{ (B)} \times p \text{ (C)} = (40/100) \times (50/100) \times (60/100) = \mathbf{12\%}$.

2.2 EXPECTED MONETARY VALUE

Main Concept

Expected Monetary Value (EMV) is a tool to quantify the risk to a project in terms of cost. Quantitative risk analysis is critical to the planning process because it helps the Project Manager evaluate the potential costs of risk events and account for them in the cost baseline using reserves.

EMV is calculated by multiplying the probability of a risk event occurring with the impact of that risk on cost. Risks as you know can be categorized as good risks (opportunities) or bad risks (threats). The impact of an opportunity is a *benefit,* and the impact of a threat is a *cost.*

The relationship between risks, their impact, and EMV is illustrated in Fig. 2.5.

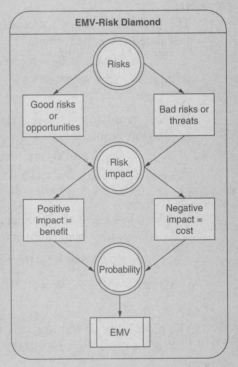

Figure 2.5 The relationship between risk, their impact, and EMV.

The sum total of EMVs for all risk events helps quantify the amount of *contingency reserve* that is warranted for a project. EMV is almost always used with the so-called "decision tree" technique to make informed decisions that take into account project risks, their impact, and their probability.

EMV can be either positive or negative:

- **Positive EMV** A positive EMV indicates that the risk actually has a positive impact or benefit if it occurs. This kind of risk is an opportunity that should be exploited and enhanced.
- **Negative EMV** A negative EMV indicates that the risk actually has a negative impact or cost if it occurs. This kind of risk is a threat that can be accepted, avoided, mitigated, or transferred. Risk avoidance or elimination is usually the preferred alternative, but it is not always practical in the real world to avoid or eliminate all bad risks or threats.

A positive EMV is denoted by a plus sign (+) prefix, and a negative EMV is denoted by a minus sign (−) prefix.

Determining the EMV allows a Project Manager to gauge whether a project has a reasonable cost-benefit ratio. But the EMV need not always be positive for a business to undertake a project. At times business or market needs may impel a company to pursue a project with a negative EMV. Of course, knowing the EMV gives the Project Manager the opportunity to present facts to senior management and to align the selection of the project with the company's goals.

TIP: *EMV is used to quantify overall project risk, not individual risk events.*

Expected Monetary Value Problems on the PMP Exam

TYPES OF PROBLEMS TO EXPECT

You can expect two types of EMV problems in the PMP exam:

(1) Computing the EMV given both the probability and the value of all risk events or outcomes.

(2) Computing the EMV given the probability and value of one out-come and the EMV of the other. Remember not to get confused when you do not see the probability of one of the outcomes. In this case you have to assume that the risk has a 100% probability of occurring.

DIFFICULTY LEVEL

You don't need a math background to score a home run on EMV prob-lems. An understanding of the question being asked and simple applica-tion of the formula will do it.

NUMBER OF PROBLEMS TO EXPECT

Expect at least one problem (or possibly two) regarding EMV on the exam.

FORMULAS

You need to know only one formula for calculating EMV on the PMP exam.

$$EMV = impact \times probability$$

INSIDER TIPS

- The EMV problems on the PMP exam may not explicitly state whether the cost impact of a risk is positive or negative. So read each problem carefully to make this determination.
- The EMV can never be greater than the value of the impact (since the probability is limited to a value between 0 and 1).

SAMPLE SOLVED PROBLEMS

1. If a new business project has an 80% chance of earning $50,000 and a 15% chance of losing $25,000, what is the expected EMV of that project?

Step 1: Find the EMV of the positive outcome.

Recall that $EMV = impact \times probability$

So in this case, that would be $50,000 \times (80/100) = 40,000$.

Because this risk has a positive impact on the project, the EMV is shown as positive, with a plus sign: (+) $40,000.

Step 2: Find the EMV of the negative outcome.

Recall that EMV = impact $\times$ probability

So in this case, that would be $(-25,000) \times (15/100) = -3,750$.

Because this risk has a negative impact on the project, the EMV is shown as negative, with a minus sign: (–) $3,750.

Step 3: Calculate the EMV of the project.

From Step 1, you have (+) 40,000.

From Step 2, you have (–) $3,750.

Thus the combined EMV for the project is $[40,000 + (-3,750)] =$ (+) $36,250.

2. The Work Breakdown Structure (WBS) of a certain project includes three work packages: A, B, and C. What is the expected EMV of that project given the following data?

Work Package A has a schedule slippage risk probability of 20% and an impact of $20,000.

Work Package B has a cost overrun risk probability of 50% and an impact of $30,000.

Work Package C has a resource attrition risk probability of 15% and an impact of $15,000.

Step 1: Find the EMV of Work Package A.

Recall that EMV = value $\times$ probability

So EMV (Package A) = $(20/100) \times (-20,000) = (-)$4,000$.

Step 2: Find the EMV of Work Package B.

Recall that EMV = value $\times$ probability

So EMV (Package B) = $(50/100) \times (-30,000) = (-)$ $15,000.

Step 3: Find the EMV of Work Package C.

Recall that EMV = value $\times$ probability

So EMV (Package C) = $(15/100) \times (-15,000) = (-)$ $2,250.

Step 4: Find the EMV of all three work packages.

From Steps 1, 2, and 3:

EMV = $(-4,000) + (-15,000) + (-2,250) = (-)$ $21,250.

3. If a project has the following risk assessments, calculate the EMV:

(i) 30% probability of schedule overrun. Impact of $30,000

(ii) 20% probability of cost overrun. Impact of $10,000

(iii) A new technology that could save $20,000

(iv) Defect rework costing $10,000 to fix

Step 1: Find the EMV of Risk 1.

EMV (Risk 1) = (30/100) × (−30,000) = −$9,000.

Step 2: Find the EMV of Risk 2.

EMV (Risk 2) = (20/100) × (−10,000) = −$2,000.

Step 3: Find the EMV of Risk 3.

EMV (Risk 3) = (100/100) × 20,000 = $20,000.

Step 4: Find the EMV of Risk 4.

EMV (Risk 4) = (100/100) × (−10,000) = −$10,000.

Step 5: Calculate EMV of all four risks.

EMV (all risks) = (−9,000) + (−2,000) + 20,000 + (−10,000)

EMV = (−) $1,000.

EXERCISE PROBLEMS

1. If a new business project has an 80% chance of earning $20,000 and a 15% chance of losing $25,000, what is the expected EMV of that project?

   A. (+) $16,000
   B. (+) $3,750
   C. (−) $12,250
   D. (+) $12,250

2. A new business project has a 15% chance of losing $25,000, but there is new technology that can save $20,000 in the project's costs. What is the expected EMV of that project?

   A. (+) $20,000
   B. (−) $3,750
   C. (+) $16,250
   D. (−) $16,250

3. A business project has a 10% chance of resource attrition with an impact of $10,000, but there is new technology that can save $1,000 in the project's costs. What is the expected EMV of that project?

A. $0
B. (−) $1,000
C. (+) $1,000
D. (+) $2,000

4. A new business project includes work packages A, B, and C. What is the expected EMV of that project given the following:

Work Package A has a schedule overrun risk probability of 10% with an impact of $10,000.

Work Package B has a cost underrun risk probability of 25% with an impact of $10,000.

Work Package C has a new resource learning curve risk impact of $15,000.

A. (−) $13,500
B. (−) $5,000
C. (+) $5,000
D. (+) $7,250

5. The WBS of a new business project includes work packages A, B, and C. What is the expected EMV of that project given the following data:

Work Package A has a schedule risk that might cost $10,000.

Work Package B has a cost risk probability of 25% with an impact of $10,000.

Work Package C has a positive impact of $30,000.

A. (−) $17,500
B. (−) $10,000
C. (+) $10,000
D. (+) $17,500

6. If a project has the following risk assessments, calculate the EMV.

(i)   20% probability of schedule overrun. Impact of $30,000
(ii) 25% probability of cost overrun. Impact of $10,000
(iii) A new technology that could save $10,000
(iv) Defect rework costing $5,000 to fix

A. (−) $13,500
B. (−) $3,500
C. (+) $3,500
D. (+) $13,500

7. If a project has the following risk assessments, calculate the EMV.

   1. Impact of $10,000 based on new government regulations
   2. 30% probability of cost overruns resulting in a loss of $30,000
   3. Workflow adjustments that could save 20 hours work netting $10,000
   4. Defect rework costing $4,000 to fix

   A. (−) $23,000
   B. (−) $13,000
   C. (+) $13,000
   D. (+) $23,000

8. A company is trying to determine if it should build a system in-house or outsource it. The setup cost for in-house is $20,000 and the cost of failure is $2,000. A design change can save $5,000. What is the EMV of doing the project in-house?

   A. (−) $23,000
   B. (−) $17,000
   C. (+) $17,000
   D. (+) $23,000

9. A company decides to skip prototyping for a project. The cost of design failure could be $30,000, but the total cost that will be saved by not prototyping is $40,000. What is the EMV of that project?

   A. (−) $10,000
   B. (+) $10,000
   C. (+) $30,000
   D. (+) $40,000

10. A company is evaluating whether it needs to do regression testing for a project because the project is already behind schedule. Doing functional testing would save $30,000 worth of effort by the testers and only delay the project by 10 days causing loss of revenue

of $10,000. Not doing the testing would violate the testing clauses of the contract costing the company $50,000. What is the EMV of the decision?

A. (−) $30,000
B. (+) $10,000
C. (+) $30,000
D. (+) $40,000

Exercise Answers

1. **D**

   Recall the EMV formula: EMV = impact × probability.

   Because the new business has an 80% chance of earning $20,000, this is a positive risk.

   $20,000 × 80/100 = (+) \$16,000$

   A 15% chance of losing $25,000 constitutes a negative risk.

   $25,000 × 15/100 = (−) \$3,750$

   Total EMV = 16,000 + (−) 3,750 = (+) **$12,250.**

2. **C**

   Recall the EMV formula: EMV = impact × probability.

   Since the new business has a 15% chance of losing $25,000, this is a negative risk.

   $= 25,000 × 15/100 = (−) \$3,750$

   The technology change that can save $20,000 is a positive risk.

   Total EMV = 20,000 + (−) 3,750 = (+) **$16,250.**

3. **A**

   Recall the EMV formula: EMV = impact × probability.

   A 10% chance of resource attrition causing an impact of $10,000 is a negative risk.

   $= 10,000 × 10/100 = (−) \$1,000$

   The technology change that can save $1,000 is a positive risk.

   Total EMV = 1,000 + (−) 1,000 = **$0.**

4. **A**

   Recall the EMV formula: EMV = impact × probability.

   Work Package A has a schedule overrun risk probability and is a negative risk.

   10,000 × 10/100 = (–) $1,000

   Work Package B has a cost underrun risk probability and is a positive risk.

   10,000 × 25/100 = (+) $2,500

   Work Package C has a new resource learning curve risk impact of $15,000. This is a negative risk.

   Total EMV = (–) 1,000 + 2,500 + (–) 15,000

   Total EMV = (–) **$13,500.**

5. **D**

   Recall the EMV formula: EMV = impact × probability.

   Work Package A is a negative risk costing $10,000.

   Work Package B has a cost risk probability of 25% and an impact of $10,000.

   10,000 × 25/100 = (–) $2,500

   Work Package C has a positive impact of $30,000.

   Total EMV = (–) 10,000 + (–) 2,500 + 30,000

   Total EMV = (+) **$17,500.**

6. **B**

   Schedule overrun implies a negative risk and the EMV is 30,000 × 20/100 = (–) $6,000.

   Cost overrun also implies a negative risk and the EMV is 10,000 × 25/100 = (–) $2,500.

   Technology saving is a positive risk of (+) $10,000.

   Defect rework is negative risk of $5,000.

   Total EMV of the project is (–) 6,000 + (–) 2,500 + 10,000 + (–) 5,000 = (–) **$3,500.**

7. **B**

   Government regulations have a negative impact of (–) $10,000.

   Cost overruns are negative and the EMV = 30,000 × 30/100 = (–) $9,000.

   Resource amount saved = $10,000.

   Defect rework costing $4,000 to fix is a negative risk.

   Total EMV = (–) 10,000 + (–) 9,000 + 10,000 + (–) 4,000 = (–) **$13,000.**

8. **B**

   The setup cost for in-house is $20,000 and the cost of failure is $2,000. So the total negative risk is (–) $22,000. A design change can save $5000. So the total EMV will be

   (–) 22,000 + 5,000 = (–) **$17,000.**

9. **B**

   The cost of design failure could be $30,000 and the total cost that will be saved is $40,000 by not prototyping. So the total EMV is

   (–) 30,000 + 40,000 = (+) **$10,000.**

10. **A**

    Doing functional testing would save $30,000. The loss of revenue of $10,000 is a negative risk. So is the lack of testing that will cost the company $50,000. So the total EMV is

    30,000 + (–) 10,000 + (–) 50,000 = (–) **$30,000.**

2.3 DECISION TREE

Main Concept

Decision tree analysis is a tool used to model decisions using the impact and probability of outcomes of the decision and selecting the best possible course of action. It is usually used in conjunction with the EMV (discussed in Section 2.2) to graphically represent the outcomes of decisions and their EMV in a tree-like structure. To be effective, the users of decision trees should have knowledge about the possible outcomes of a decision and how likely each is to occur (i.e., their probability).

A decision tree typically consists of "decision nodes," "chance nodes," and "end nodes" as depicted in Fig. 2.6.

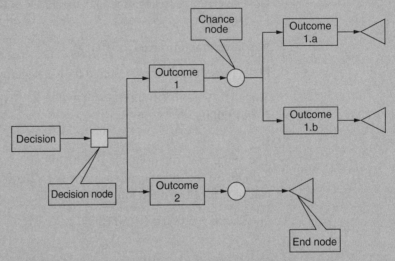

Figure 2.6 Decision tree.

The decision node (square) represents a decision to be made, and the branches out of this decision node represent its possible outcomes. Chance nodes (circles) depict the probability of the occurrence of an outcome. End nodes (triangles) represent the final value/impact of an outcome.

The PMP exam does not require you to draw a decision tree, but familiarity with the concepts discussed above is crucial.

TIP: *Most people are visual. A decision tree can be used for other business decisions even when EMV is not taken into account.*

Decision Tree Problems on the PMP Exam

TYPES OF PROBLEMS TO EXPECT

There is only one type of decision tree problem on the PMP exam:

(1) Deciding which decision is the best course of action by calculating the EMV of all the possible outcomes.

DIFFICULTY LEVEL

Probability and EMV are the fundamental concepts being tested in this kind of problem. So if you have meticulously studied those sections and worked through the exercise problems, decision trees should be easy for you.

NUMBER OF PROBLEMS TO EXPECT

Expect at least one problem if not two regarding a decision tree on the PMP exam.

FORMULAS

There is only one type of decision tree problem on the PMP exam. The formula for solving it is the same as for solving EMV problems:

$$EMV = value \times probability$$

INSIDER TIPS

- The typical decision tree problem on the PMP exam has only a simplified version of a decision tree, usually with only two outcomes.

SAMPLE SOLVED PROBLEMS

1. ABC Company is faced with a make-or-buy decision. Based on the following diagram, what is the EMV of the buy decision?

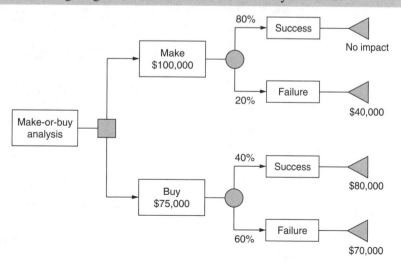

Step 1: Find the EMV of the buy outcome—success.

Recall that EMV = value × probability.

So in this case, the EMV of the buy/success outcome is $75,000 + (40/100) × 80,000

= 75,000 + 32,000

= $107,000

Step 2: Find the EMV of the buy outcome—failure.

Recall that EMV = value × probability.

So in this case, the EMV of the buy/failure outcome is $75,000 + 70,000

= (−) $145,000 (recall that because this outcome has a negative impact on the project, the EMV is shown with a minus sign.)

Step 3: Calculate the EMV of the buy decision.

From Step 1, you have (+) 107,000.

From Step 2, you have (−) 145,000.

Thus the EMV is (−) $138,000.

EXERCISE PROBLEMS

1. ABC Company is facing a build or upgrade decision. Based on the following diagram, which one should the company choose, and what would be the EMV of that decision?

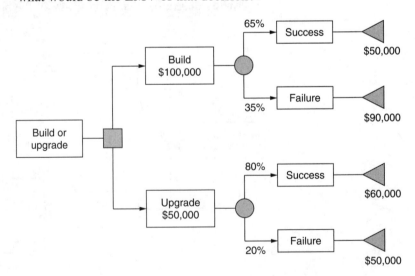

A. Build with an EMV of $101,000
B. Build with an EMV of $164,000
C. Upgrade with an EMV of $88,000
D. Upgrade with an EMV of $164,000

2. A Project Manager needs to take a flight to visit a client. He has a choice of Airline A or Airline B. Based on the following diagram, which one should he choose, and what would be the EMV of that decision?

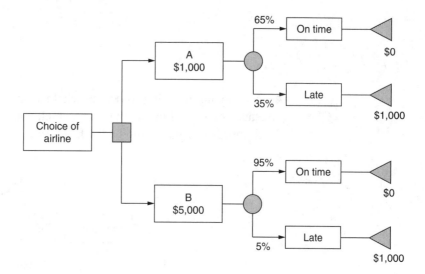

A. Airline A with an EMV of $650
B. Airline A with an EMV of $4,950
C. Airline B with an EMV of $1,350
D. Airline B with an EMV of $4,950

3. A company is trying to decide between System A and System B. Based on the following diagram, which one should it choose, and what would be the EMV of that decision?

A. System A with an EMV of $62,500
B. System A with an EMV of $10,250
C. System B with an EMV of $62,500
D. System B with an EMV of $10,250

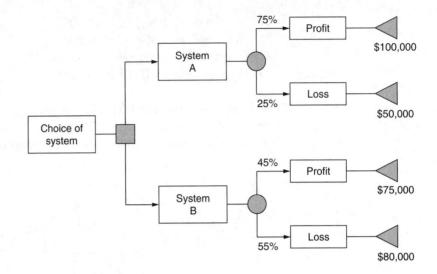

4. A company is trying to decide whether or not to regression test. Based on the following diagram, which course of action should it choose, and what would be the EMV of that decision?

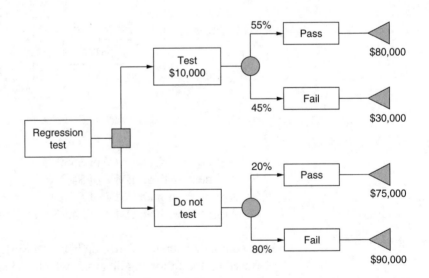

A. Test with an EMV of $40,500
B. Test with an EMV of (–) $57,000
C. Don't test with an EMV of $40,500
D. Don't test with an EMV of (–) $57,000

5. You as a Project Manager have to choose between Vendor A and Vendor B for a project. Based on the following diagram, which one should you choose, and what would be the EMV of your decision?

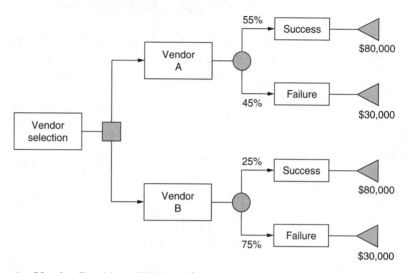

A. Vendor B with an EMV of $30,500
B. Vendor B with an EMV of (−) $2500
C. Vendor A with an EMV of $30,500
D. Vendor A with an EMV of (−) $2500

6. ABC Company is facing a make or buy decision. Based on the following diagram, which course of action should it choose, and what would be the EMV of that decision?

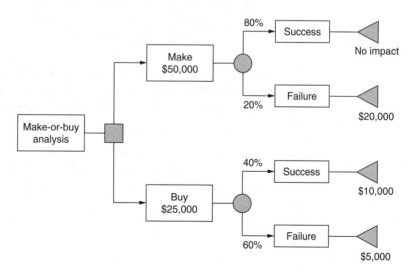

A. Make with an EMV of $46,000
B. Make with an EMV of $26,000
C. Buy with an EMV of $46,000
D. Buy with an EMV of $26,000

7. A Project Manager needs to choose between two competing projects A and B. Based on the following diagram, which one should she choose, and what would be the EMV of her decision?

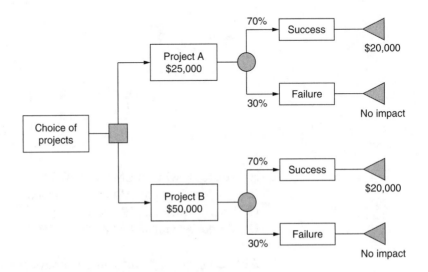

A. Project A with an EMV of $39,000
B. Project A with an EMV of $64,000
C. Project B with an EMV of $39,000
D. Project B with an EMV of $64,000

8. A company is trying to decide between purchasing Tool A or Tool B. Based on the following diagram, which one should it buy, and what would be the EMV of that decision?

A. Tool B with an EMV of $52,500
B. Tool A with an EMV of (–) $750
C. Tool A with an EMV of $52,500
D. Tool B with an EMV of (–) $750

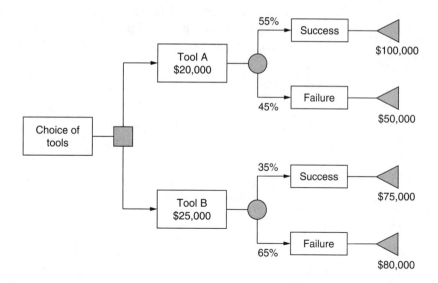

9. A company is trying to decide between purchasing Software A or Software B to augment security. Based on the following diagram, which one should it buy, and what would be the EMV of that decision?

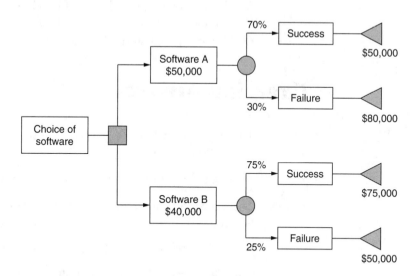

A. Software A with an EMV of $83,750
B. Software A with an EMV of $61,000
C. Software B with an EMV of $61,000
D. Software B with an EMV of $83,750

10. You as a Project Manager have to select between Vendor A and Vendor B for a project. Based on the following diagram, which one should you choose, and what would be the EMV of your decision?

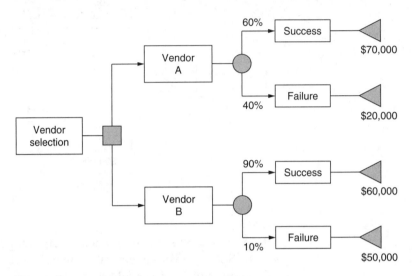

A. Vendor B with an EMV of $34,000
B. Vendor B with an EMV of $49,000
C. Vendor A with an EMV of $49,000
D. Vendor A with an EMV of $34,000

Exercise Answers

1. A

The decision to build = $100,000 + 65% of $50,000 + 35% of (− $90,000)

$100,000 + 65% of $50,000 + 35% of (− $90,000) = $101,000

The decision to upgrade = $50,000 + 80% of $60,000 + 20% of (− $50,000)

$50,000 + 80% of $60,000 + 20% of (− $50,000) = $88,000

The decision EMV is **$101,000** (the larger of $101,000 and $88,000).

2. D

Choosing Airline A = $1,000 + 65% of $0 + 35% of (− $1,000)

= $1,000 + 0 − 350 = $650

Choosing Airline B = $5,000 + 95% of $0 + 5% of (− $1,000)

= $5,000 + 0 − 50 = $4,950

The Decision EMV is **$4,950** (the larger of $650 and $4,950).

3. **A**

Choosing System A = 75% of $100,000 + 25% of (− $50,000) = $62,500

Choosing System B = 45% of $75,000 + 55% of (− $80,000) = (−) $10,250

The Decision EMV is **$62,500** (the larger of $62,500 and (−) $10,250).

4. **A**

Deciding to test = 10,000 + 55% of $80,000 + 45% of (− $30,000) = $40,500

Deciding not to test = 20% of $75,000 + 80% of (− $90,000) = (−) $57,000

The Decision EMV is **$40,500** (the larger of $40,500 and (−) $57,000).

5. **C**

Choosing Vendor A = 55% of $80,000 + 45% of (− $30,000) = $30,500

Choosing Vendor B = 25% of $80,000 + 75% of (− $30,000) = (−) $2,500

The decision EMV is **$30,500** (the larger of $30,500 and (−) $2,500).

6. **A**

EMV of a make decision = 50,000 + 80% of $0 − 20% of ($20,000) = $46,000

EMV of an buy decision = 25,000 + 40% of $10,000 + 60% of (−) $5000 = $26,000

The decision EMV is **$46,000** (the larger of $46,000 and $26,000).

7. **D**

EMV of choosing Project A = 25,000 + 70% of $20,000 + 30% of $0 = $39,000

EMV of choosing Project B = 50,000 + 70% of $20,000 + 30% of $0 = $64,000

The decision EMV is **$64,000** (the larger of $39,000 and $64,000).

8. **C**

EMV of choosing Tool A = 20,000 + 55% of $100,000 + 45% of (− $50,000) = $52,500

EMV of choosing Tool B = 25,000 + 35% of $75,000 + 65% of (− $80,000) = (−) $750

The decision EMV is **$52,500** (the larger of $52,500 and (−) $750).

9. **D**

EMV of choosing Software A = 50,000 + 70% of $50,000 + 30% of (− $80,000) = $61,000

EMV of choosing Software B = 40,000 + 75% of $75,000 + 25% of (− $50,000) = $83,750

The decision EMV is **$83,750** (the larger of $83,750 and $61,000).

10. **B**

EMV of choosing Vendor A = 60% of $70,000 + 40% of (− $20,000) = $34,000

EMV of choosing Vendor B = 90% of $60,000 + 10% of (− $50,000) = $49,000

The decision EMV is **$49,000** (the larger of $34,000 and $49,000).

2.4 PROJECT COST ESTIMATION

Main Concept

Estimating project cost is one of the most important functions of a Project Manager. Apart from creating a budget for the project, it is crucial to accurately estimate and compute contingency and managerial reserves. Cost estimates also help in making the choice between multiple projects by providing a basis for comparing estimated costs to project benefits. But an estimate is only as good as the predictive skills of the estimator and the accuracy of the data being used. Estimates can be of different types: a rough order-of-magnitude (ROM) estimate or a definitive estimate.

A rough order-of-magnitude estimate is a high-level estimate that is usually used when the scope of the project is not entirely known, such as in the early stages (Initiation). It is also known as a ballpark estimate; and allows an error margin of −50% to +100%.

A definitive estimate on the other hand is more precise and accurate and reflects the cost estimate of a project with a clearly defined scope. Definitive estimates are usually formulated only after there is sufficient detail about the tasks involved in the project and the level of effort required. The precision of such an estimate should be within −5% to +10% of the expected cost.

There are also different techniques for estimating project costs with differing levels of accuracy, these are listed in the following sections.

Analogous Estimating

Analogous estimating is a top-down cost-estimating technique that is based on the estimator's experience with similar projects in the past and any available historical information. There is not much calculation involved in this method of estimation, but its accuracy does depend on the expert judgment of the Project Manager (estimator). This technique is typically used when there is a lack of detailed information about the project and all its components, or when there is a similar comparable project.

Bottom-up Estimating

In bottom-up estimating, you start out by estimating cost at the level of work-breakdown-structure (WBS) work packages. These cost estimates are then aggregated into a control account. The sum of all control accounts across the project is the project cost. The cost baseline comprises the project cost and a contingency reserve. The contingency reserve is also referred to as the cost of dealing with residual risks (i.e., risks that remain after risk response planning) or "known unknowns" (i.e., risks that are identified and have been accepted). A management reserve is added to account for "unknown unknowns," which are basically unanticipated issues and risks (i.e., risks that have not been identified). The addition of the management reserve to the cost baseline gives the total cost budget for the project. The concept of cost aggregation is illustrated in the "cost pyramid" shown in Fig. 2.7.

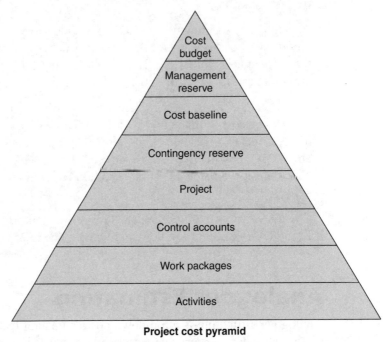

Project cost pyramid

Figure 2.7 Project cost pyramid.

Parametric Estimating

Parametric cost estimation is a statistical model that estimates the cost of the project based on specific project characteristics. Parametric estimates work best when they are applied to projects that are similar to projects you have worked on in the past.

Types of Costs

As a Project Manager, you need to know about these estimating techniques, and you also need to know how to classify the different types of costs that can be attributed to a project. These types are listed below.

- **Fixed Costs** Fixed costs are costs that do not vary depending on sales or production. Examples of fixed costs are rent and set-up costs. Fixed costs should always be included in project cost estimates.

- **Variable Costs**   Variable costs vary depending on sales or production. Variable costs include the costs of materials, supplies, and wages.
- **Direct Costs**   Direct costs are costs that are directly related to a project. Examples include team travel, team wages, recognition, awards, and cost of materials.
- **Indirect Costs**   Indirect costs are costs that are not directly related to the project. Examples include taxes, goodwill, and cleaning services. These costs can be divided across multiple projects.

Project Cost Estimation Problems on the PMP Exam

TYPES OF PROBLEMS TO EXPECT

There are only two formulas to remember for estimating project costs. However, you can expect to see four types of cost estimating problems on the PMP exam:

(1) Computing the total cost of the project
(2) Computing cost ranges while estimating rough order of magnitude
(3) Computing the definitive estimate
(4) Computing cost aggregation

DIFFICULTY LEVEL

Yes! Finally this is one section where you can kick off your shoes and relax. All you need is the ability to add.

NUMBER OF PROBLEMS TO EXPECT

On the PMP exam, expect one problem on cost estimation.

FORMULAS

Cost baseline = project cost + contingency reserve
Cost budget = cost baseline + management reserve

INSIDER TIPS

- Remembering if the management reserve is included in the cost baseline or the cost budget can be confusing for some people. An easy way to remember is to recall that budgets are most relevant to management, and thus the cost budget includes the management reserve.

SAMPLE SOLVED PROBLEMS

1. Listed below are the costs for each of the phases of the software development life cycle (SDLC).

Phase	Cost (in dollars)
Analysis	$40,000
Design	$80,000
Implementation	$120,000
Testing and monitoring	$60,000
Maintenance	$20,000

What is the project life-cycle cost estimate?

Step 1: Compute the life-cycle cost.

Life-cycle cost is the cost of all the stages in the life cycle put together.

$= 40,000 + 80,000 + 120,000 + 60,000 + 20,000$

$= \$320,000$

2. A project has 5 work packages and each of them is estimated to cost $40,000. The contingency reserve is $50,000 and there is a managerial reserve of $20,000. What is the cost budget for this project?

Step 1: Compute the project cost.

Because there are 5 work packages and each of them is estimated to cost $40,000, the total cost is $\$40,000 \times 5 = \$200,000$.

Step 2: Compute the cost baseline.

Cost baseline = project cost + contingency reserve

Cost baseline = $200,000 + 50,000 = \$250,000$

Step 3: Compute the cost budget.

Cost budget = cost baseline + management reserve

Cost budget = $250,000 + 20,000 = \$270,000$

3. In the initiation phase, a Project Manager estimates the cost of a project to be $500,000. He/she further estimates $40,000 as a contingency reserve and $60,000 as a management reserve. What is the rough order of magnitude estimate?

Step 1: Compute the cost baseline.

Cost baseline = project cost + contingency reserve

Cost baseline = 500,000 + 40,000 = $540,000

Step 2: Compute the cost budget.

Cost budget = cost baseline + management reserve

Cost budget = 540,000 + 60,000

Cost budget = $600,000

Step 3: Compute the rough order of magnitude estimate.

The rough order of magnitude estimate is in the range of −50% to +100% from actual. So in this case, the range is from $300,000 to $1,200,000.

4. In the initiation phase, a Project Manager estimates the cost of a project to be $500,000. She/he further estimates $40,000 as a contingency reserve and $60,000 as a management reserve. What is the definitive estimate?

Step 1: Compute the cost baseline.

Cost baseline = project cost + contingency reserve

Cost baseline = 500,000 + 40,000 = $540,000

Step 2: Compute the cost budget.

Cost budget = cost baseline + management reserve

Cost budget = 540,000 + 60,000

Cost budget = $600,000

Step 3: Compute the definitive estimate.

The definitive estimate is in the range of −5% to +10% from actual, so in this case, the range is from $570,000 to $660,000.

EXERCISE PROBLEMS

1. Listed below are the costs for each of the phases of the software development life cycle.

Phase	Cost (in dollars)
Analysis	$40,000
Design	$60,000

Development	$200,000
Quality assurance	$60,000
Implementation	$30,000
Closing & maintenance	$20,000

What is the total project life-cycle cost estimate?

A. $390,000
B. $400,000
C. $410,000
D. $420,000

2. Listed below are the costs for each of the phases of the software development life cycle.

Phase	Cost (in dollars)
Initiation	$20,000
Planning	$30,000
Executing	$100,000
Implementation	$30,000
Closing	$20,000

The company reserves 10% as a contingency. What is the total project life-cycle cost estimate?

A. $200,000
B. $220,000
C. $225,000
D. $400,000

3. A project has 10 work packages, and each one is estimated to cost $40,000. The contingency reserve is $50,000 and there is a managerial reserve of $70,000. What is the cost budget for this project?

A. $130,000
B. $160,000
C. $400,000
D. $520,000

4. The total estimated cost of a project is $200,000. The contingency reserve is $10,000 and there is a managerial reserve of $20,000. What is the cost budget for this project?

A. $50,000
B. $200,000
C. $210,000
D. $230,000

5. In the initiation phase, a Project Manager estimates the cost of a project to be $200,000. He/she further estimates $20,000 as a contingency reserve and $30,000 as a management reserve. What is the rough order of magnitude estimate?

A. $125,000 to $500,000
B. $250,000 to $500,000
C. $125,000 to $250,000
D. $125,000 to $700,000

6. The cost baseline for a project is estimated to be $220,000. The management reserve is set at $30,000. What is the definitive estimate of the cost budget?

A. $125,000 to $500,000
B. $250,000 to $500,000
C. $237,500 to $275,000
D. $125,000 to $700,000

7. The budget of the new online inventory tracking system is $650,000; which includes $80,000 in contingency reserves and $70,000 in management reserves. What is the rough order of magnitude estimate of the project?

A. $125,000 to $500,000
B. $325,000 to $650,000
C. $325,000 to $1,250,000
D. $325,000 to $1,300,000

8. The cost baseline for a web based banking application is $480,000. This includes a contingency estimate of $80,000. The management reserve is estimated at $120,000. What is the definitive estimate?

A. $125,000 to $700,000
B. $500,000 to $700,000
C. $570,000 to $660,000
D. $540,000 to $700,000

9. What is the range for a rough order of magnitude estimate?

    A. −10% to +15%
    B. −75% to +100%
    C. −5% to +10%
    D. −50% to +100%

10. What is the range for a definitive estimate?

    A. −50% to +100%
    B. −75% to +100%
    C. −10% to +15%
    D. −5% to +10%

Exercise Answers

1. **C**

    The total project life-cycle cost estimate is

    $40,000 + $60,000 + $200,000 + $60,000 + $30,000 + $20,000

    = **$410,000.**

2. **B**

    The total project life-cycle estimate is

    $20,000 + $30,000 + $100,000 + $30,000 + $20,000

    = $200,000

    Don't forget to add the 10% contingency expense, which is $20,000 (10% of $200,000).

    200,000 + 20,000 = **$220,000.**

3. **D**

    First, calculate the project cost. The project cost is $40,000 × 10 = $400,000 because there are 10 modules.

    To calculate the cost baseline, use the formula:

    Cost baseline = project cost + contingency reserve

    Cost baseline = 400,000 + 50,000 = $450,000

Cost budget = cost baseline + management reserve

Cost budget = 450,000 + 70,000 = **$520,000.**

4. **D**

First, calculate the project cost. The project cost is $20,000 × 10 = $200,000 because there are 10 modules.

To calculate the cost baseline, use the formula:

Cost baseline = project cost + contingency reserve

Cost baseline = 200,000 + 10,000 = $210,000

Cost budget = cost baseline + management reserve

Cost budget = 210,000 + 20,000 = **$230,000.**

5. **A**

To calculate the cost baseline, use the formula:

Cost baseline = project cost + contingency reserve

Cost baseline = 200,000 + 20,000 = $220,000

Cost budget = cost baseline + management reserve

Cost budget = 220,000 + 30,000 = $250,000

A rough order of magnitude estimate implies a range of −50% to + 100%, or **$125,000** to **$500,000.**

6. **C**

To calculate the cost budget, use the formula:

Cost budget = cost baseline + management reserve

Cost budget = 220,000 + 30,000 = $250,000

A definitive estimate implies a range of −5% to +10%, or

$237,500 to **$275,000.**

7. **D**

Cost budget = $650,000

A rough order of magnitude estimate implies a range of −50% to + 100%, or **$325,000** to **$1,300,000.**

8. **C**

   To calculate the cost baseline, use the formula:

   Cost baseline = project cost + contingency reserve

   Cost baseline = 400,000 + 80,000 = $480,000

   Cost budget = cost baseline + management reserve

   Cost budget = 480,000 + 120,000 = $600,000

   A definitive estimate implies a range of −5% to +10%, or **$570,000 to $660,000.**

9. **D**

   A rough order of magnitude estimate implies a range of **−50%** to **+100%.**

10. **D**

    A definitive estimate implies a range of **−5%** to **+10%.**

2.5 CRITICAL PATH AND FLOAT

Main Concept

Once a project's deliverables have been broken down into manageable pieces by use of a WBS, and activities have been defined and estimated, it is time to focus on schedule development. A useful tool for scheduling is the network diagram, which depicts the relationships and dependencies between project activities.

In a network diagram, the longest path is the critical path, which indicates the shortest time required to complete the project. Every time someone asks how long the project will take, the answer is always the length of the critical path.

There is always at least one critical path in a network diagram. The path is critical because a delay in any of the activities along this path will most certainly delay the project too. So a Project Manager should focus on controlling and managing these critical activities. Multiple critical paths can exist, but this just means there is more for the Project Manager to manage, more potential failure points (delays), and hence more risk to the project.

The critical path is also the path along which activities have "zero float." *Float* is the length of time a particular activity can be delayed without affecting the completion time of the project as a whole. It is also referred to as *slack*. Activities that are on the critical path have no breathing room; they all have "zero float."

There are several different types of floats, as follows:

- **Total Float**   This is the length of time an activity can be delayed without affecting the following milestone or the project end date. The word *float* is often used to mean "total float."
- **Free Float**   This is the length of time an activity can be delayed without affecting the start date of the successive activities.
- **Project Float**   This is the length of time a project can be delayed without impacting the externally imposed project completion date.
- **Negative Float**   One important thing to remember is that float can be negative. Negative float indicates that the planned duration of the activity or project exceeds the time available, which also means that you are behind schedule. In such a case, the negative float is the length of time that needs to be made up in order for the project or activity to meet its deadline and finish on schedule.

Network Diagramming Methods

Before delving into sample math problems involving float and critical path duration calculations, let's look at some ways to represent a network diagram.

- **Precedence Diagramming Method (PDM) or Activity on Node Method**
  In this diagramming method, nodes denote activities and arrows connecting nodes denote precedence (dependencies). The relationship between the activities can take the following forms:
  - **Start to Start**   One activity must start before another activity can start. For example, suppose a software development team is following agile methodology. In this case, the development effort begins as soon as the task of gathering requirements begins.
  - **Start to Finish**   One activity must start before another activity can finish. An example would be the activity of conducting code reviews in a software development project should begin prior to completion of all coding activity. Doing so helps constantly monitoring the quality of code and minimizes redesign.
  - **Finish to Start**   One activity must finish before another activity can start. For example, the development effort on a project using the waterfall method needs to be completed before testing can begin.

- **Finish to Finish** One activity must finish before another activity can finish. For example, a testing task cannot be finished until the task of fixing all defects is complete.
- **Arrow Diagramming Method (ADM) or Activity on Arrow Method** In this diagramming method, arrows denote activities, with the tail indicating the start of the activity and the head representing the end. Common practice is to have a node at each end of the arrow. This kind of diagram can be used to illustrate a sequence of activities. There can only be one type of relationship between activities in this kind of diagram:
  - **Finish to Start** One activity must finish before another activity can start. For example, on a project using the waterfall approach, the activity of gathering requirements must finish before development can begin.

Critical Path and Float Calculation Methods: CPM, PERT, and GERT

The mathematical techniques most widely used by Project Managers to calculate critical path and float are the following:

- **Critical Path Method (CPM)** In this method, the first task is to create a network diagram using either the PDM or the ADM described above and identifying the critical path(s). The network path(s) closest in duration to the critical path is called the "near-critical path." There can be multiple critical paths and near-critical paths in a single network diagram. The greater the number of critical paths and near-critical paths, the higher the risk to the project's schedule. Once the critical path has been identified, the float or slack of noncritical tasks can be calculated. The float in these activities can be used to compress the schedule of the critical path activities using methods such as crashing and fast tracking. In a network diagram, the notation used to represent an activity along with its start and end dates, duration, and float/slack is shown in Fig. 2.8.

Early start	Duration	Early finish
	Task name	
Late start	Slack	Late finish

Figure 2.8 Notation used to represent an activity or task.

Float is calculated as

$$\text{Float} = \text{Late start} - \text{Early start or Late finish} - \text{Early finish}$$
$$= \text{LS} - \text{ES or LF} - \text{EF}$$

- **Program Evaluation and Review Technique (PERT)**   In PERT analysis, optimistic, pessimistic, and most likely activity time span estimates are assigned weights and then averaged to arrive at a realistic date of completion. This technique will be discussed in detail in Sec. 2.7 of this chapter.
- **Graphical Evaluation and Review Technique (GERT)**   The rarely used GERT is similar to the PERT, but it also accounts for looping (i.e., activities that are performed multiple times). Examples of such activities are the task of fixing defects and the task of testing them. These tasks are iterative in nature.

Critical Path and Float Problems on the PMP Exam

TYPES OF PROBLEMS TO EXPECT

On the PMP exam, you can expect three types of problems involving critical path and float:

(1) Computing the critical path
(2) Computing the float
(3) Computing the effect of change on the critical path

DIFFICULTY LEVEL

Once you practice the sample problems and diligently do the exercise problems, these problems should be easy for you.

NUMBER OF PROBLEMS TO EXPECT

Expect 5 to 10 problems regarding critical path. Usually 5 problems are based on a single network diagram. Expect a maximum of 2 such problems.

FORMULAS

The only formula that you need to remember is the formula for calculating float/slack:

$$\text{Float/slack} = \text{Late start} - \text{Early start}$$
$$= \text{LS} - \text{ES}$$

or

$$\text{Float/slack} = \text{Late finish} - \text{Early finish}$$
$$= LF - EF$$

INSIDER TIPS

• Always double-check your critical path and float. Refer to your diagrams at all times to make sure you do not overlook anything.

SAMPLE SOLVED PROBLEMS

1. Given the information below, answer the questions that follow after drawing a network diagram.

Task/Activity	Preceding Activity	Duration in days
Task A	Start	5 days
Task B	Task A	4 days
Task C	Task B	6 days
Task D	Task B	2 days
Task E	Task C	7 days
Task F	Task C, Task D	1 day
Task G	Task E, Task F	5 days

Step 1: Create the network diagram.

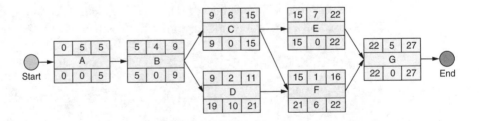

The critical path is the longest path in the network diagram. After listing all the possible paths in the network diagram, you need to choose the path that has the greatest number of days. Listed below are the different paths and the number of days required for each one. Based on the preceding diagram, the different paths are

Path	Duration (days)
Start – Task A – Task B – Task C – Task E – Task G – End	27
Start – Task A – Task B – Task D – Task F – Task G – End	17
Start – Task A – Task B – Task C – Task F – Task G – End	21

Thus the critical path is Start – Task A – Task B – Task C – Task E – Task G – End, which takes 27 days.

Step 2: Answer the questions.

a. If the duration of Task C is increased to 10 days, what will be the effect on the project?

First determine if Task C is in the critical path. Because it is, it will make a difference to the duration of the project. Now the different paths are

Path	Duration
Start – Task A – Task B – Task C – Task E – Task G – End	31
Start – Task A – Task B – Task D – Task F – Task G – End	17
Start – Task A – Task B – Task C – Task F – Task G – End	25

The critical path "Start – Task A – Task B – Task C – Task E – Task G – End" now has duration of 31 days. So the project duration is extended by 4 days.

b. If the duration of Task C is decreased to 2 days, what will be the effect on the project?

First determine if Task C is in the critical path. Because it is, it will make a difference to the duration of the project. Now the different paths are

Path	Duration
Start – Task A – Task B – Task C – Task E – Task G – End	23
Start – Task A – Task B – Task D – Task F – Task G – End	17
Start – Task A – Task B – Task C – Task F – Task G – End	17

The critical path "Start – Task A – Task B – Task C – Task E – Task G – End" now has duration of 23 days. So the project duration is compressed by 4 days.

c. If the duration of Task F is increased to 3 days, what will be the effect on the project?

First determine if Task F is in the critical path. Because it is not, it will not make a difference to the duration of the project. Now the different paths are

Path	Duration
Start – Task A – Task B – Task C – Task E – Task G – End	27
Start – Task A – Task B – Task D – Task F – Task G – End	19
Start – Task A – Task B – Task C – Task F – Task G – End	23

The critical path is still "Start – Task A – Task B – Task C – Task E – Task G – End" with a duration of 27 days. Task F is not on the critical path, so the project duration is not altered.

d. If the duration of Task F is increased to 7 days, what will be the effect on the project?

First determine the paths and the duration:

Path	Duration
Start – Task A – Task B – Task C – Task E – Task G – End	27
Start – Task A – Task B – Task D – Task F – Task G – End	23
Start – Task A – Task B – Task C – Task F – Task G – End	27

There are now two critical paths: "Start – Task A – Task B – Task C – Task E – Task G – End" and "Start – Task A – Task B – Task C – Task E – Task G – End," each with a duration of 27 days. It is acceptable to have multiple critical paths in a project. But the risk element increases with the increase in the number of critical paths.

2. Given the information below, answer the questions that follow after drawing a network diagram.

Activity	Predecessor	Duration in days
A	Start	3
B	Start	6
C	A, B	12
D	B	5
E	D	4
F	C	6
G	E, F	8
H	E	7
I	H	3

Step 1: Create the network diagram.

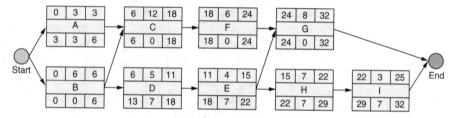

Step 2: Answer the questions.

a. What is the critical path?

The critical path is the longest path in the network diagram. After listing all the possible paths in the network diagram, you need to choose the path that has the most number of days. Listed below are the different paths and the number of days required for each one. Using the precedence diagramming method, you can chart the different paths as follows:

Path	Duration
Start – Task A – Task C – Task F – Task G – End	29 days
Start – Task B– Task D – Task E – Task H – Task I – End	25 days
Start – Task B– Task D – Task E – Task G – End	23 days
Start – Task B – Task C – Task F – Task G – End	32 days

The critical path is "Start – B – C – F – G – End" with duration of 32 days.

b. What is the duration of the critical path?

The duration of the critical path is 32 days.

c. What is the float of Activity B?

Referring to the network diagram, you can see that Activity B is a part of the critical path with a float of 0.

d. What is the float of Activity H?

Referring to the network diagram, you can see that Activity H is not a part of the critical path. Its float is $LS - ES = 22 - 15 = 7$ (or $LF - EF = 29 - 22 = 7$).

EXERCISE PROBLEMS

1. A project has the following specifications: Gathering requirements (Task A) takes 10 days beginning at the start of the project. Once the requirements are gathered, documenting them (Task B) takes 5 days to complete. After documenting, the team must create a design document (Task C) which takes 10 days. Once the design document is created, the team starts development efforts (Task D), which take 25 days. At the same time, the team also starts creating test cases (Task E), which takes 7 days. After tasks D and E are completed, testing efforts (Task F) will commence and take a total of 10 days. If Task C takes 15 days instead of 10, which of the following is true?

   A. The critical path takes 60 days.
   B. The critical path takes 0 days.
   C. The critical path increases by 5 days.
   D. The critical path changes to 55 days.

2. Based on the information below, answer the questions that follow using a network diagram.

   • Task A takes 25 days and can begin as soon as the project starts.
   • Task B needs to be done after the completion of Task A and takes an estimated 20 days.
   • Task C can start when Task B starts and takes an estimated 30 days.
   • Task D can start as soon as Task B is complete and will take at least 10 days to complete.
   • Task E can start as soon as tasks C and D are completed and takes 40 days to complete.

a. What is the duration of the critical path?

    A. 95 days
    B. 45 days
    C. 55 days
    D. 0 days

b. What is the float of Task C?

    A. 25 days
    B. 15 days
    C. 5 days
    D. 0 days

c. If another Task F with a duration of 5 days is scheduled to be completed after Task C, but before starting Task E, what is the new critical path duration?

    A. 120 days
    B. 100 days
    C. 55 days
    D. 0 days

d. What is the float of Activity B?

    A. 10 days
    B. 5 days
    C. 0 days
    D. −5 days

3. Based on the information below, answer the questions that follow using a network diagram.

Activity	Estimated Duration (weeks)
Start – Activity A	4
Start – Activity B	2
Activity A – Activity C	5
Activity B – Activity C	2
Activity B – Activity E	4
Activity C – Activity D	4
Activity C – Activity E	2
Activity E – Activity F	1
Activity D – End	3
Activity F – End	3

a. What is the critical path?

    A. Start – A – C – E – F – End
    B. Start – A – C – D – End
    C. Start – B – C – D – End
    D. Start – B – E – F – End

b. What is the duration of the critical path?

    A. 10 weeks
    B. 11 weeks
    C. 15 weeks
    D. 16 weeks

c. If Task E is eliminated, how many weeks of the project can be saved?

    A. 0 weeks
    B. 1 week
    C. 2 weeks
    D. 4 weeks

d. What is the near critical path?

    A. Start – A – C – E – F – End
    B. Start – A – C – D – End
    C. Start – B – C – D – End
    D. Start – B – E – F – End

4. Use the network diagram below to answer the questions that follow.

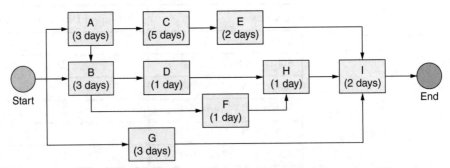

a. What is the duration of the critical path in this network?

    A. 5 days
    B. 9 days
    C. 12 days
    D. 15 days

b. What is the float for Activity F?

A. 1 day

B. 3 days

C. 5 days

D. 7 days

c. If the customer insists that the project be finished by day 10 without any modifications to the activity durations or start times, what would be the new float of Activity F?

A. −3 days

B. −1 day

C. 1 day

D. 3 days

5. Given the portion of the network diagram illustrated below, what is the late finish (LF) of Activity C?

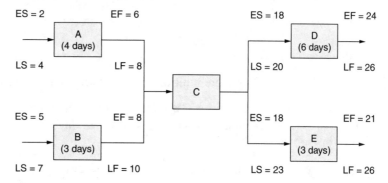

A. 9

B. 23

C. 19

D. 22

Exercise Answers

1. C

Using the given information, you can construct the following network diagram:

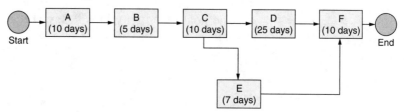

The critical path in this diagram is Start – A – B – C – D – F – End with a duration of 60 days. Because Task C is on the critical path, if it takes 15 days to complete instead of 10, then project (i.e. critical path) duration increases by **5 days** (choice C).

2. Using the given information, you can construct the following network diagram:

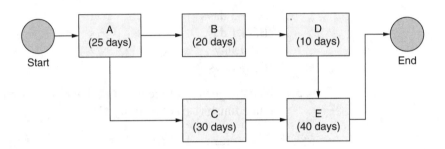

There are multiple critical paths in this diagram:

- Start – A – B – D – E – End = 95 days
- Start – A – C – E – End = 95 days

a. A

Critical path duration = **95 days.**

b. D

The float of Task C is **0 days** because it is on the critical path.

c. B

Inserting a new Task F to be completed after completion of Task C and prior to initiating Task E increases the duration of the critical path to **100 days.** It also changes the network diagram so that there is now only 1 critical path. The former other critical path becomes the near critical path.

d. C

The float of Activity D is **0 days** because it is on the critical path.

3. Using the given information, you can construct the following network diagram:

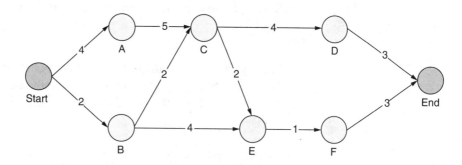

a. B

The possible paths in the network and their lengths are as follows:

- Start – A – C – D – End = 16 weeks
- Start – A – C – E – F – End = 15 weeks
- Start – B – E – F – End = 10 weeks
- Start – B – C – D – End = 11 weeks
- Start – B – C – E – F – End = 10 weeks

Therefore, the critical path is **"Start – A – C – D – End."**

b. D

The duration of the critical path is **16 weeks.**

c. A

Because **Activity E is not on the critical path,** eliminating it will not reduce the project duration and thus will not save any time.

d. A

The near critical path is the path that is closest in length to the critical path. From the paths listed in the answer to question (a) above, you can see that the near critical path is **"Start – A – C – E – F – End"** with duration of 15 weeks.

4. In the network diagram shown, the critical path is "Start – A – C – E – I – End."

   a. C

   Based on the network diagram, the critical path duration is **12 days.**

   b. C

   Modify the network diagram as shown below to calculate the float of each activity. Based on the diagram, the float of Activity F is LF – EF = 9 – 4 = **5 days** (also LS – ES = 8 – 3 = 5 days)

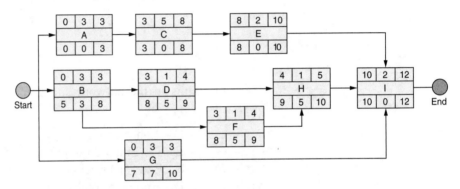

   c. D

   The newly imposed finish date of day 10 becomes the LF for Activity I. Using that as your reference, you can recalculate the LF and LS for the activities in the network diagram and modify the diagram as shown below.

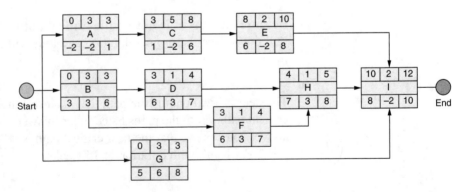

   Based on this new modified network diagram, the float for Activity F is **3 days.**

5. C

Working backward through the network and determining the LS for the tasks or activities immediately following Task C can determine the LF for a task.

In this case the two tasks following Task C are Task D and Task E, with LS of 20 and 23, respectively. Task C will need to finish before both tasks D and E start; that is, Task C will need to finish before the LS of both tasks D and E.

If you take the smaller LS of Task D (20) and subtract 1 from it, you will see that Task C will need to finish at the latest by day 19.

Therefore, the LF of **Task C = 19.**

2.6 CRASHING

Main Concept

Once the network diagram has been created, it's time to establish the project schedule. Often the original schedule needs to be compressed because of customer demands, budgetary concerns, or other market forces. Schedule compression is a tricky proposition because it almost always involves a trade-off between cost, time, and performance. The objective of schedule compression is to shorten the project schedule without changing the scope, in order to meet project goals, objectives, and deadlines. Typical compression strategies include adding more resources, improving processes, and similar measures. Some of the more common schedule compression techniques are discussed in the following sections.

Fast Tracking

Fast tracking involves working on tasks in parallel rather than in sequence, as originally scheduled. Fast tracking can also be achieved by overlapping tasks such that a subsequent task begins before the current task completes. Parallel execution of tasks and their potential overlap increases risk and the potential for rework. If the increased risk can be addressed effectively, then fast tracking is a viable option to compress a schedule. Fast tracking problems are not part of the PMP exam and thus are not discussed here.

Crashing

Crashing is a technique used to determine the optimal trade-off between costs and schedule that can be used to compress a project's schedule at minimal cost. Crashing almost always adds to the project cost, usually in terms of additional resource requirements.

Crashing Problems on the PMP Exam

TYPES OF PROBLEMS TO EXPECT

There is only one type of crashing problem to expect on the PMP exam:

(1) Find the tasks that are candidates for crashing and the order in which to crash them.

DIFFICULTY LEVEL

Crashing is not math-oriented. Once you know how to identify critical-path tasks that have potential for crashing, it is just a question of ordering those tasks based on their crashing cost.

NUMBER OF PROBLEMS TO EXPECT

Expect a maximum of one problem on crashing on the PMP exam.

FORMULAS

There are no formulas to remember for problems involving the crashing technique.

INSIDER TIPS

- Make sure you determine the critical path before you look at options for crashing. If a particular task is not on the critical path, there is no point in spending extra cash to crash it.

SAMPLE SOLVED PROBLEMS

1. Assume the following network diagram:

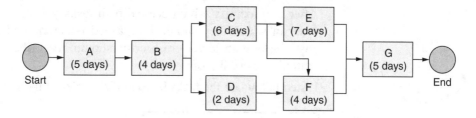

	Time Required (Days)		Cost ($)		Extra Crashing Cost Per Saved Day ($)
Activity	Normal	Crash	Normal	Crash	
	Col A	Col B	Col C	Col D	Col D – Col C/(Col A – Col B)
A	5	2	10,000	12,000	666.67
B	4	2	15,000	22,500	3,750
C	6	1	4,500	8,000	700
D	2	1	6,000	12,000	6,000
E	7	3	34,000	40,000	1,500
F	4	2	14,000	20,000	3,000
G	5	2	5,000	30,000	8,333.34

Which of the activities would you crash and in what order?

Step 1: Find the critical path.

Using the precedence diagramming method, you can chart the different paths as follows:

Start – Task A – Task B – Task C – Task E – Task G – End	27 days
Start – Task A – Task B – Task D – Task F – Task G – End	20 days
Start – Task A – Task B – Task C – Task F – Task G – End	24 days

The critical path is "Start – Task A – Task B – Task C – Task E – Task G – End" with a duration of 27 days. So now you know that these are the tasks that you need to crash.

Step 2: Find the order in which to crash costs.

After you identify which critical-path tasks you need to crash, the order in which they should be crashed is determined by the crash cost. Tasks with lower crash costs should be crashed before those with higher crash costs.

Listed below are the tasks in ascending order of their crash cost.

Task	Crash Cost ($)
A	666.67
C	700
E	1,500
B	3,750
G	8333.34

Thus the crash order should be tasks A, C, E, B, G.

EXERCISE PROBLEMS

1. Consider activities A, B, and D that are on the critical path. Activities E and C are in the near-critical path.

Task	Normal Time (weeks)	Crash Time (weeks)	Normal Cost ($)	Crash Cost ($)
A	5	3	12,000	14,000
B	6	3	14,000	20,000
C	4	2	16,000	18,000
D	5	3	15,000	25,000
E	4	2	11,000	12,000

Which of the following lists the tasks in the correct crashing order?

A. Tasks C, B, D
B. Tasks B, D, A
C. Tasks A, B, D
D. Tasks C, E, A

2. Activities A, B, C, D, E, and F are on the critical path.

Task	Normal Time (weeks)	Crash Time (weeks)	Normal Cost ($)	Crash Cost ($)
A	10	5	14,000	22,000
B	15	5	20,000	26,000
C	12	8	28,000	34,000
D	20	10	26,000	36,000
E	14	10	22,000	42,000
F	16	6	13,000	33,000

Which of the following lists the tasks in the correct crashing order ?

A. Tasks B, D, C, A, F, E
B. Tasks B, C, D, A, E, F
C. Tasks B, A, D, A, F, E
D. Tasks E, F, A, C, D, B

3. Consider the following network diagram:

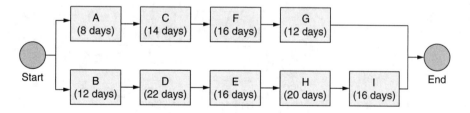

Some of these tasks can be crashed, but crashing involves extra costs, as shown in the following table:

Activity	Time Required (Days)		Cost ($)		Extra Crashing Cost Per Saved Day ($)
	Normal	Crash	Normal	Crash	
A	8	6	14,000	22,000	4,000
B	12	6	20,000	26,000	1,000
C	14	10	30,000	32,000	500
D	22	14	26,000	36,000	1,250
E	16	12	22,000	42,000	5,000
F	16	6	13,000	33,000	2,000
G	12	8	22,000	28,000	1,500
H	20	15	8,000	12,000	800
I	16	12	16,000	26,000	2,500

Which of the following lists the tasks that can be crashed in the correct order?

A. Tasks B, D, E, H, I
B. Tasks H, B, D, I, E
C. Tasks B, A, D
D. Tasks A, C, F, G

4. Consider the following network diagram:

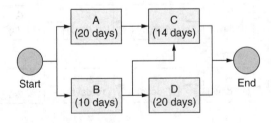

Some of these tasks can be crashed, but crashing involves extra costs, as shown in the following table:

Activity	Time Required (Days) Normal	Crash	Cost ($) Normal	Crash	Extra Crashing Cost Per Saved Day ($)
A	20	15	8,000	12,000	800
B	10	5	20,000	22,000	400
C	14	10	30,000	32,000	500
D	20	16	26,000	30,000	1,000

Which of the following lists the tasks that can be crashed in the correct order?

A. Tasks B, C, A, D
B. Tasks B, D, C
C. Tasks A, C
D. Tasks C, A

5. A certain project has a float of −6 months. Given the following information, which activities would you crash and in what order?

Activity	Normal Time (months)	Crash Time (months)	Time Saved (months)	Normal Cost ($)	Crash Cost ($)	Extra Cost for Crashing ($)	Extra Crashing Cost Per Saved Month ($)
A	8	6	2	14,000	22,000	8,000	4,000
B	12	6	6	20,000	26,000	6,000	1,000
C	14	10	4	30,000	32,000	2,000	500
D	22	14	8	26,000	36,000	10,000	1,250
E	16	12	4	22,000	42,000	20,000	5,000

A. Task B
B. Tasks A, C
C. Tasks A, E
D. Tasks D, E

6. A certain project has a float of −5 months. Given the following information, which activities would you crash and in what order?

Activity	Normal Time (months)	Crash Time (months)	Time Saved (months)	Normal Cost ($)	Crash Cost ($)	Extra Cost for Crashing ($)	Extra Cost Per Saved Month ($)
A	8	6	2	16,000	22,000	6,000	3,000
B	12	9	3	20,000	26,000	6,000	2,000
C	14	12	2	30,000	38,000	8,000	4,000
D	23	20	3	30,000	33,000	3,000	1,000
E	16	11	5	22,000	39,000	17,000	3,400

A. Tasks A and B
B. Task E
C. Tasks A and D
D. Tasks C and D

7. Activities A, B, and E are on the critical path. Activities D and C are on the near-critical path.

Task	Normal Time (weeks)	Crash Time (weeks)	Normal Cost ($)	Crash Cost ($)
Task A	8	6	16,000	22,000
Task B	12	9	20,000	26,000
Task C	14	12	30,000	38,000
Task D	23	20	30,000	33,000
Task E	16	11	22,000	39,000

Which of the following lists the tasks that can be crashed in the correct crashing order?

A. Tasks B, A, E
B. Tasks A, B, E
C. Tasks A, B, C, D
D. Tasks A, B, C, D, E

8. Assume the following network diagram:

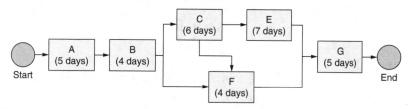

Activity	Time Required (Days)		Cost ($)		Extra Crashing Cost Per Saved Day ($)
	Normal	Crash	Normal	Crash	
	Col A	Col B	Col C	Col D	Col D – Col C/(Col A – Col B)
A	5	2	10,000	10,600	200
B	4	2	15,000	15,650	325
C	6	1	8,000	8,500	100
E	7	3	40,000	41,000	250
F	4	2	20,000	20,750	375
G	5	2	5,000	5,900	300

Which of the following lists the tasks that can be crashed in the correct crashing order?

A. Tasks A, B, C

B. Tasks A, B, C, E

C. Tasks A, B, C, E, G

D. Tasks C, A, E, G, B

9. A certain project has a float of −5 months. Given the following information, which activities would you crash and in what order?

Activity	Normal Time (months)	Crash Time (months)	Time Saved (months)	Normal Cost ($)	Crash Cost ($)	Extra Cost for Crashing ($)	Extra Cost Per Saved Month ($)
A	5	2	3	10,000	10,600	600	200
B	4	2	2	15,000	15,650	650	325
C	6	1	5	8,000	12,000	4,000	800
D	5	2	3	5,000	5,900	900	300
E	4	2	2	20,000	20,750	750	375

A. Tasks A, B
B. Task C
C. Tasks A, E
D. Tasks D, E

10. Assume the following network diagram:

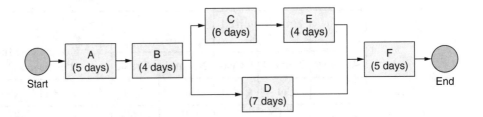

	Time Required (Days)		Cost ($)		Extra Crashing Cost Per Saved Day ($)
Activity	Normal	Crash	Normal	Crash	
	Col A	Col B	Col C	Col D	Col D – Col C/(Col A – Col B)
A	5	2	25,000	28,000	1000
B	4	2	30,000	36,000	3000
C	6	1	45,000	52,000	1400
D	7	3	40,000	52,000	3000
E	4	2	60,000	72,000	6000
F	5	2	21,000	25,500	1500

Which of the following lists the tasks that can be crashed in the correct crashing order?

A. Tasks A, B, C, E, F
B. Tasks A, C, F, B
C. Tasks A, C, F
D. Tasks A, C, F, B, E

Exercise Answers

1. **C**

   In this case, the extra crashing cost per saved week has not been given. So you need to compute that amount.

   Because only tasks A, B, and D are in the critical path, you need to compute the crashing cost for only those activities.

   To compute the extra crashing cost per saved week, first compute the number of saved weeks = normal time – crash time.

   Second, find the difference between the crash cost and normal cost.

   Third, divide the difference in cost by the difference in weeks, as shown in the table below.

Activity	Time Required (Weeks)		Cost ($)		Extra Crashing Cost Per Saved Week ($)
	Normal	Crash	Normal	Crash	
	Col A	Col B	Col C	Col D	Col D – Col C/(Col A – Col B)
A	5	3	12,000	14,000	= 14,000 – 12,000/(5 – 3) = $1,000 per week
B	6	3	14,000	20,000	= 20,000 – 14,000/(6 – 3) = $2,000 per week
D	5	3	15,000	25,000	= 25,000 – 15,000/(5 – 3) = $5,000 per week

   The order in which the tasks should be crashed is **A, B, D.**

2. **A**

   Because all the activities are in the critical path, you need to compute the crashing cost for all of them, as shown in the following table:

Activity	Time Required (Weeks)		Cost ($)		Extra Crashing Cost Per Saved Week ($)
	Normal	Crash	Normal	Crash	
	Col A	Col B	Col C	Col D	Col D – Col C/(Col A – Col B)
Task A	10	5	14,000	22,000	= 22,000 – 14,000/(10 – 5) = $1,600 per week
Task B	15	5	20,000	26,000	= 26,000 – 20,000/(15 – 5) = $600 per week
Task C	12	8	28,000	34,000	= 34,000 – 28,000/(12 – 8) = $1,500 per week
Task D	20	10	26,000	36,000	= 36,000 – 26,000/(20 – 10) = $1,000 per week
Task E	14	10	22,000	42,000	= 42,000 – 22,000/(14 – 10) = $5,000 per week
Task F	16	6	13,000	33,000	= 33,000 – 13,000/(16 – 6) = $2,000 per week

The order in which the tasks should be crashed is **B, D, C, A, F, E**

3. **B**

First, find the critical path: B, D, E, H, I.

Because you are given the extra crashing cost per saved day for these tasks, all you need to do is arrange the tasks on the critical path in ascending order of their crashing cost.

The correct order is **H, B, D, I, E.**

4. **D**

First, find the critical path: A, C.

Because you are given the extra crashing cost per saved day for these tasks, all you need to do is arrange the tasks on the critical path in ascending order of their crashing cost.

The correct order is **C, A.**

5. **A**

The different combinations to save 6 months are as follows:

Tasks	Cost ($)
Task B	1,000
Tasks A and C	4,500
Tasks A and E	9,000

The task that has the least cost is **Task B,** so that should be the task to crash.

6. **B**

Tasks	Cost ($)
Task E	3,400
Tasks A and B	5,000
Tasks A and D	4,000
Tasks C and B	6,000
Tasks D and C	5,000

The task that has the least cost is **Task E,** so that should be the task to crash.

7. **A**

Because tasks A, B, and E are on the critical path, you need to compute the extra crashing cost per saved week for each of them.

Activity	Time Required (Weeks)		Cost ($)		Extra Crashing Cost Per Saved Week ($)
	Normal	Crash	Normal	Crash	
	Col A	Col B	Col C	Col D	Col D – Col C/(Col A – Col B)
Task A	8	6	16,000	22,000	= 22,000 – 16,000/(8 – 6) = $3,000 per week
Task B	12	9	20,000	26,000	= 26,000 – 20,000/(12 – 9) = $2,000 per week
Task E	16	11	22,000	39,000	= 39,000 – 22,000/(16 – 11) = $3,400 per week

The crashing order should be **B, A, E.**

8. **D**

   In the network diagram, the critical path is "Start – A – B – C – E – G – End." Because you are given the extra crashing cost per saved day, all you need to do is arrange the tasks on the critical path in ascending order of their crashing cost savings: **C, A, E, G, B.**

9. **A**

   The following tasks or task combinations can each save 5 months of time.

Tasks	Cost ($)
Task C	800
Tasks A and B	525
Tasks A and E	575
Tasks D and B	625
Tasks D and E	675

   The combination that has the least cost is **Tasks A and B.** Therefore, they are the ones that should be crashed.

10. **D**

    The critical path is A, B, C, E, and F. The crash order should be in ascending order of the crash cost per week, that is **A, C, F, B, E.**

2.7 PERT, STANDARD DEVIATION, AND VARIANCE

Main Concept

A big part of the Project Manager's role is to work with his or her team to estimate the time required to complete project activities. The Project Manager is not necessarily the best judge of how much time each person will take. Studies suggest that time estimates are best sought from the person actually doing the job. However, too often the person giving the estimate can be overly optimistic or pessimistic about his or her abilities and timeframes. Program Evaluation and Review Technique (PERT) is a tool that seeks to work around these limitations, addresses the uncertainty in estimations of project parameters, and uses a weighted average to gauge the amount of time each activity is expected to take to complete. The weighted average takes into account three levels of estimates for each activity: an "optimistic" estimate, a "pessimistic" estimate, and a "most likely" estimate.

The expected activity duration times calculated with PERT can also be used to estimate the total expected completion time of activities on the critical path. But simply adding up those expected activity durations is not enough. The total time estimate needs to be accompanied by an indication of the extent of variation from the PERT-calculated critical-path duration. This variation is illustrated by the variance and standard deviation of activities on the critical path. In mathematical terms, the standard deviation is also referred to as "Sigma" (σ). You can calculate the total standard deviation of activities on the critical path by finding the square root of the sum of the variances of the activities on the path. (There is a statistical inaccuracy that occurs by just adding standard deviations, so the correct approach is to add the variances and then find the square root of the sum.) You can also use PERT to generate a "what if" scenario for time estimates.

The formulas for finding expected completion times, variance, and standard deviation are discussed below.

TIP: *Microsoft Project provides for the ability to perform quantitative risk analysis using the PERT model. You can find a brief discussion of this approach in App. B of the book.*

PERT, Standard Deviation, and Variance Problems on the PMP Exam

TYPES OF PROBLEMS TO EXPECT

On the PMP exam, you can expect two types of problems involving PERT estimates:

(1) Computing the PERT
(2) Computing the range of estimate

DIFFICULTY LEVEL

These problems are relatively simple if you remember the formulas.

NUMBER OF PROBLEMS TO EXPECT

Expect one or two problems regarding PERT estimates on the PMP exam.

FORMULAS

Assume

> O is the optimistic estimate
> M is the most likely estimate
> P is the pessimistic estimate

Then,

- PERT expected time/duration $(E) = (O + 4M + P)/6$
- Standard deviation $= (P - O)/6$
- Variance $= [(P - O)/6]^2$

INSIDER TIPS

- In PERT, the optimistic and pessimistic estimates are each assigned a weight of 1. The "most likely" estimate is assigned a weight of 4. These weights can be changed based on the experience of the team member providing the estimates.
- Do not add standard deviations! Only variances of tasks can be added. Then find the square root of the total variance to determine the total standard deviation.

SAMPLE SOLVED PROBLEMS

1. You are planning to use PERT for scheduling your project. One project task has an optimistic schedule of 22 days and a pessimistic schedule of 40 days, but is most likely to be completed in 35 days. What is the expected completion time for the task?

   Step 1: Find the expected duration using PERT.

   Recall the formula for PERT: $E = (O + 4M + P)/6$

   So all you have to do is substitute the values into the formula:

   $E = [22 + (35 \times 4) + 40]/6$

   $E = 33.67$ days

2. You are the Project Manager for a software development project. You have gathered time estimates for the design, development, and QA modules of the project as shown in the table below. Given this data, what is the expected duration of the project and the range of variation for this estimate?

Module	Optimistic (days)	Most Likely (days)	Pessimistic (days)
Design	12	18	25
Development	10	15	20
QA	4	6	10

Step 1: Find the expected duration, variance, and standard deviation for each task using PERT.

The easiest way to do the math here is to use the tabular format to find PERT, standard deviation, and variance.

Module	O (days)	M (days)	P (days)	PERT	Std deviation	Variance
				$(O+4M+P)/6$	$(P-O)/6$	$[(P-O)/6]\,2$
Development	12	18	25	18.17	2.17	4.69
QA	10	15	20	15	1.67	2.78
Implementation	4	6	10	6.33	1	1

Step 2: Compute expected duration and variance for the whole project.

The sum of the durations of all tasks is the duration of the project.

Duration = 18.17 + 15 + 6.33 = 39.5 days

The sum of the variances of all tasks is the total project variance.

Variance = 4.69 + 2.78 + 1 = 8.47 days

Step 3: Calculate the standard deviation for the project.

The standard deviation for the project is the square root of the total project variance.

Standard deviation. = $\sqrt{(8.47)}$ = 2.91 days

The standard deviation indicates the range of variation in the calculated project duration.

EXERCISE PROBLEMS

1. The optimistic schedule estimate for a project is 10 days, the pessimistic schedule estimate is 20 days, and most likely estimate is 15 days. What is the PERT?

   A. 10 days
   B. 15 days
   C. 16 days
   D. 20 days

2. The optimistic schedule estimate for a project is 4 days, the pessimistic schedule estimate is 12 days, and most likely schedule estimate is 8 days. What is the PERT?

   A. 2 days
   B. 5 days
   C. 6 days
   D. 8 days

3. Assume a project has the following tasks and time estimates. What is the variance for this project?

Module	Optimistic (days)	Most Likely (days)	Pessimistic (days)
Login	22	27	30
Registration	14	10	16
Automation	20	25	30

   A. 0.11 days
   B. 2.16 days
   C. 2.78 days
   D. 4.67 days

4. Assume a project has the following tasks and time estimates. What is the variance for this project?

Module	Optimistic (days)	Most Likely (days)	Pessimistic (days)
Login	12	18	25
Registration	22	27	30
Automation	14	20	24

A. 1.78 days
B. 2.79 days
C. 4.71 days
D. 9.27 days

5. Assume a project has the following tasks and time estimates. What is the standard deviation for this project?

Module	Optimistic (days)	Most Likely (days)	Pessimistic (days)
Login	24	37	40
Registration	21	24	28
Automation	25	30	35

A. 1.37 days
B. 2.13 days
C. 3.36 days
D. 11.29 days

6. Assume a project has the following tasks and time estimates. What is the standard deviation for this project?

Module	Optimistic (days)	Most Likely (days)	Pessimistic (days)
Login	16	18	25
Registration	24	30	36
Automation	30	35	41

A. 2.25 days
B. 4 days
C. 3.10 days
D. 9.6 days

7. Assume a project has the following tasks and time estimates. What is the range of estimate for this project assuming 1 sigma?

Module	Optimistic (days)	Most Likely (days)	Pessimistic (days)
A	22	27	30
B	14	20	24
C	4	6	10

A. Ranges between 50 and 53 days
B. Ranges between 50 and 56 days
C. Ranges between 53 and 56 days
D. Ranges between 60 and 66 days

8. Assume the optimistic time estimate for a project is 10 days, the pessimistic estimate is 20 days, and the most likely estimate is 15 days. The project uses 3 sigma estimates. Using PERT, what is the range of this estimate?

A. Ranges between 10 and 15 days
B. Ranges between 10 and 20 days
C. Ranges between 15 and 20 days
D. Ranges between 13 and 16 days

9. Assume the optimistic time estimate for a project is 16 days, the pessimistic estimate is 34 days, and the most likely estimate is 24 days. Using PERT, what is the range of this estimate and the type of distribution?

A. Ranges between 22 and 28 days with a bell distribution
B. Ranges between 22 and 28 days with a normal distribution
C. Ranges between 22 and 28 days with a uniform distribution
D. Ranges between 22 and 28 days with a beta distribution

10. A Project Manager is given an optimistic time estimate of 22 days for a task. The pessimistic estimate is 32 days and the most likely estimate is 26 days. If he is using PERT, what range of estimate will he report and which type of distribution will he use?

A. Ranges between 25 and 29 days with a normal distribution
B. Ranges between 25 and 29 days with a normal distribution

C. Ranges between 25 and 29 days with a uniform distribution

D. Ranges between 25 and 29 days with a beta distribution

Exercise Answers

1. **B**

   Recall that PERT = [P + (4 × M) + O]/6

   So by substituting values, you get [20 + (4 × 15) + 10]/6 = **15 days.**

2. **D**

   This again is a straightforward calculation of PERT.

   Recall that PERT = [P + (4 × M) + O]/6.

   So by substituting values, you get, [12 + (4 × 8) + 4]/6 = **8 days.**

3. **D**

   The easiest way to solve this problem is to extend the table to include variance.

Module	Optimistic (days)	Most Likely (days)	Pessimistic (days)	Variance [(P − O)/6]^2
Login	22	27	30	1.78*
Registration	14	10	16	0.11*
Automation	20	25	30	2.78*

   * These numbers have been rounded to 2 decimal places.

   The total project variance =1.78 + 0.11 + 2.78 = **4.67 days.**

4. **D**

Module	Optimistic (days)	Most Likely (days)	Pessimistic (days)	Variance [(P − O)/6]^2
Login	12	18	25	4.69*
Registration	22	25	30	1.78*
Automation	14	20	24	2.79*

   * These numbers have been rounded to 2 decimal places.

   The total variance = 4.69 + 1.78 + 2.78 = **9.25 days.**

5. **C**

Remember that you cannot add standard deviations of different tasks. You need to find the variance of all tasks and then find the standard deviation.

Module	Optimistic (days)	Most Likely (days)	Pessimistic (days)	Variance [(P – O)/6]^2
Login	24	37	40	7.13*
Registration	21	24	28	1.37*
Automation	25	30	35	2.79*

* These numbers have been rounded to 2 decimal places.

The total variance = 7.13 + 1.37 + 2.79 = 11.29

Now, finding the standard deviation is a one-step process.

Recall that standard deviation = square root of the total variance = $\sqrt{11.29}$ = **3.36 days.**

6. **C**

Module	Optimistic (days)	Most Likely (days)	Pessimistic (days)	Variance [(P – O)/6]^2
Login	16	18	25	2.25
Registration	24	30	36	4
Automation	30	35	41	3.35*

* This number has been rounded to 2 decimal places.

The total variance = 2.25 + 4 + 3.35 = 9.6

Now, finding the standard deviation is a one-step process.

Recall that standard deviation = square root of the total variance = $\sqrt{9.64}$ = **3.10 days.**

7. **B**

Module	Optimistic (days)	Most Likely (days)	Pessimistic (days)	PERT (O + 4M + P)/6	Variance [(P – O)/6]^2
A	22	27	30	26.67	1.77
B	14	20	24	19.67	2.79
C	4	6	10	6.33	1

The total variance = 1.77 + 2.79 + 1 = 5.56

Now, finding the standard deviation is a one-step process.

Recall that standard deviation = square root of the total variance = $\sqrt{5.56} = 2.36$.

That is approximately 3 days.

Note: 2.36 is taken as 3 days since this estimates number of man days.

Total number of days for the project is 26.67 + 19.67 + 6.34 = 52.68

So the range is 53 days ± 3 days or **50 to 56 days.**

8. **B**

First calculate the PERT using the formula [P + (4 × M) + O]/6 = [20 + (4 × 15) + 10]/6 = 90/6 = 15 days.

Now, find the standard deviation using the formula (P − O)/6 = (20 −10)/6 = 10/6 = 1.67.

Because you need to compute 3 sigma, you have to get 1.67 × 3 = 5

So the estimate for this project is 15 days ± 5 days or **10 to 20 days** uniform distribution.

9. **B**

First calculate the PERT using the formula [P + (4 × M) + O]/6 = [34 + (4 × 24) + 16]/6 = 146/6 = 24.34 days.

Now, find the standard deviation using the formula (P − O)/6 = (34 − 16)/6 = 18/6 = 3.

The estimate for this project is 25 days ±3 days or **22 to 28 days.**

A normal distribution appropriately indicates the variation of the data. However, a beta distribution would skew the results toward the pessimistic end of the curve.

10. **A**

First calculate the PERT using the formula [P + (4 × M) + O]/6 = [32 + (4 × 26) + 22]/6 = 158/6 = 26.34 days.

Now, find the standard deviation using the formula (P − O)/6 = (32 − 22)/6 = 10/6 = 1.67.

The estimate for this project is 27 days ± 2 days or **25 to 29 days.**

A normal distribution appropriately indicates the variation of the data. However, a beta distribution would skew the results toward the pessimistic end of the curve.

2.8 CONTRACT COST ESTIMATION

Main Concept

Business today is characterized by an ever-increasing number of outsourced projects and short-term contracts. Now more than ever, a Project Manager's talent portfolio must include vendor management skills.

To govern the legal aspect of the vendor relationship, there must be a contract. A contract is an agreement between two or more parties to exchange goods and services for a predetermined compensation. There are several contract types. A Project Manager needs to ensure that the proper type of contract is chosen, based on the needs of the project and the company. The different types of contracts are discussed in the following sections.

Fixed Price

Under fixed price (FP) contracts, the buyer pays the vendor a fixed price irrespective of the time and materials it takes to complete the contract. There are several types of fixed price contracts.

- **Fixed Price Plus Incentive Fee**   Under a fixed price plus incentive fee contract, the buyer and the seller (the vendor) predetermine the target price of the contract and a ceiling price. Any cost over and above the ceiling price is borne by the vendor. Buyer and vendor also determine a target profit, and a share ratio for how a cost overrun or savings would be split between the buyer and vendor. Finally, the buyer agrees to pay the vendor an incentive fee consisting of a percentage share of the difference between the contract's target price and the vendor's actual costs. The vendor thus has an incentive to curb the actual cost far below the target price in order to increase the amount he or she earns.

- **Fixed Price Plus Award Fee**   Fixed price contracts motivate the vendor to lower expenses in order to achieve higher profits. While this type of contract protects the buyer from the risk of cost increase, it does not protect the buyer from the risk of low quality resulting from the vendor's cost-cutting efforts. To overcome this drawback, an award fee is used to motivate the vendor to maintain or surpass certain project expectations.

Cost Reimbursable

Under a cost reimbursable (CR) contract, the buyer reimburses the vendor for all costs incurred while executing the contract. There are several types of cost reimbursable contracts.

- **Cost Plus Fixed Fee** Under this type of contract, the buyer reimburses the vendor for costs incurred while executing the contract and also pays the vendor a mutually stipulated fixed fee.
- **Cost Plus Percentage of Costs Fee** Under this type of contract, the buyer reimburses the vendor for costs incurred while executing the contract and also pays the vendor a percentage of those costs as an incentive. Since this type of a contract does not provide the vendor any incentive to control costs (in fact it is an incentive to maximize costs and hence their profit), its usage is considered poor business practice. In fact, the U.S. government even prohibits this type of a contract for any federal contracts.
- **Cost Plus Incentive Fee** Under this type of contract, the buyer and the vendor predetermine the target cost of the contract and a ceiling price. Any costs over and above the ceiling price are borne by the vendor. Buyer and vendor also determine a target profit and a share ratio for how a cost overrun or savings would be split between the buyer and vendor. Finally, the buyer agrees to pay the vendor an incentive fee consisting of a percentage share of the difference between the contract's target cost and the vendor's actual costs. The vendor thus has an incentive to curb actual costs far below the target cost in order to increase the amount he or she earns.
- **Cost Plus Award Fee** Under this type of contract, the buyer reimburses the vendor for costs incurred while executing the contract and also pays the vendor an award fee based on the buyer's evaluation of the vendor's performance against set criteria.

Time and Materials

Under a time and materials (T & M) contract, the buyer compensates the vendor for the amount of time, labor, and materials spent on the project. The cost of the work in these types of contracts is quoted as a unit price (i.e., cost per unit or hourly labor cost). A T & M contract has features of both an FP contract and a CR contract, in that the unit price is fixed and the vendor does get reimbursed for all material costs. The difference is in the amount of risk a T & M contract poses for both the buyer and seller (vendor). While FP contracts are riskier for the seller and CR contracts for the buyer, a T & M contract allows buyers and sellers to share the risk. T & M contracts are most suited to short-term projects without a defined scope or to projects prone to frequent changes. Time & Material contracts are not usually ideal and suitable for long-term projects with a fixed scope.

Contract Cost Estimation Problems on the PMP Exam

TYPES OF PROBLEMS TO EXPECT

There are five types of contract cost estimation problems that you may encounter on the PMP exam:

(1) Computing the total contract cost of a cost plus incentive fee contract without a ceiling price
(2) Computing the total contract cost of a cost plus incentive fee contract with a ceiling price
(3) Computing an incentive fee.
(4) Computing the fee adjustment when there is a cost overrun
(5) Computing the fee adjustment when there is a cost underrun

DIFFICULTY LEVEL

We rate this problem type to be of medium difficulty because most project managers do not have exposure to contract management. Once you grasp the subtle differences between different types of contracts, you will have an easier time solving these problems.

NUMBER OF PROBLEMS TO EXPECT

Expect at least one problem regarding contract costs on the PMP exam.

FORMULAS

Incentive = (target cost − actual cost) × vendor's share ratio percentage

Overhead fee = vendor's target profit + incentive

Contract cost = actual cost + overhead fee

INSIDER TIPS

- If there is a cost overrun, incentive is negative. If there is a cost underrun, incentive is positive.

SAMPLE SOLVED PROBLEMS

1. Your company (the buyer) has negotiated a cost plus incentive fee contract with a vendor. The contract has a target cost of $300,000. The vendor's target profit is set at 10%, and the share ratio between buyer and vendor is set at 80/20. The actual cost of the contract ends up being $275,000. What is the total cost of the contract to the buyer?

Step 1: Find the incentive.

Incentive = (target cost − actual cost) × vendor's share ratio percentage

Incentive = (300,000 − 275,000) × 20/100

Incentive = 25,000 × 0.2 = $5,000

Step 2: Find the overhead fee.

Overhead fee = vendor's target profit + incentive

Overhead fee = (300,000 × 10%) + 5,000 = 30,000 + 5,000

Overhead fee = $35,000

Step 3: Find the contract cost.

Contract cost = actual cost + overhead fee

Contract cost = 275,000 + 35,000

Contract cost = $310,000

2. Your company has a fixed price plus incentive fee contract with a vendor. The contract has the following details:

Target price: $600,000

Target incentive fee: $100,000

Share ratio: buyer 70% and vendor 30%

Actual cost: $650,000

Ceiling price: $725,000

What is the final price your company will have to pay?

Step 1: Find the incentive.

Because the actual cost is greater than the target cost, this indicates a cost overrun. Thus the incentive is negative, as shown below.

Incentive = (target cost − actual cost) × vendor's share ratio percentage

Incentive = (600,000 − 650,000) × 30/100

Incentive = − 50,000 × 0.3 = − $15,000

Step 2: Find the overhead fee.

Overhead fee = target fee + incentive

In this case because of the cost overrun, the incentive is negative, so the incentive will be deducted from the vendor's target fee.

Overhead fee = 100,000 + (−15,000)

Overhead fee = $85,000

Step 3: Find the contract cost.

Contract cost = actual cost + overhead fee

Contract cost = 650,000 + 85,000

Contract cost = $735,000

Step 4: Compare ceiling price with contract cost.

Ceiling price < contract cost

Because the ceiling price is less than the contract cost, the fee paid to the vendor will be limited to the ceiling price, which is $725,000. Had there been no ceiling price, the contract cost would have been $735,000 as computed in Step 3.

3. Your company has a cost plus incentive fee contract with a vendor. The target cost is $150,000. The target fee is $20,000, and the share ratio between buyer and vendor is set at 80/20. The maximum fee is $30,000 and the minimum fee is $5,000. If the actual cost is $190,000, what fee does your company have to pay the vendor?

Step 1: Find the incentive.

Because the actual cost is greater than the target cost, there is a cost overrun. Thus the incentive is negative, as shown below.

Incentive = (target cost − actual cost) × vendor's share ratio percentage

Incentive = (150,000 − 190,000) × 20/100

Incentive = −40,000 × 0.2 = −$8,000

Step 2: Find the overhead fee.

Overhead fee = target fee + incentive

In this case because of the cost overrun, the incentive is negative, so the incentive will be deducted from the vendor's target fee.

Overhead fee = 20,000 + (−8,000)

Overhead fee = $12,000 (this is the fee paid to the vendor)

4. Your company has a fixed price plus incentive fee contract with a vendor. The contract has the following details:

Target cost: $100,000

Target profit: $30,000

Target price: $130,000

Ceiling price: $170,000

Share ratio: buyer 70% and vendor 30%

Actual cost: $70,000

What is the vendor's profit?

Step 1: Find the incentive.

Since the actual cost is less than the target cost,

Incentive = (target cost – actual cost) × vendor's share ratio percentage

Incentive = (100,000 – 70,000) × 30/100

Incentive = 30,000 × 0.3 = $9,000

Step 2: Find the overhead fee.

Overhead fee = target fee + incentive

Overhead fee = 30,000 + 9,000

Overhead fee = $39,000 (this is the total fee paid to the vendor, i.e., the vendor's profit from this procurement)

5. In a fixed price plus incentive fee contract, the target cost is $100,000 and the target profit is 10%. There is a ceiling price of $130,000. The share ratio between buyer and vendor is set at 70/30. If the actual cost is $110,000, what is the final value of this procurement?

Step 1: Find the incentive.

Because the actual cost is greater than the target cost, there is a cost overrun. Thus the incentive is negative, as shown below.

Incentive = (target cost – actual cost) × vendor's share ratio percentage

Incentive = (100,000 – 110,000) × 30/100

Incentive = –10,000 × 0.3 = –$3,000

Step 2: Find the overhead fee.

In this case because of the cost overrun, the incentive is negative, so the incentive will be deducted from the seller's target fee.

Overhead fee = target fee + incentive

Overhead fee = 10,000 + (−3,000)

Overhead fee = $7,000

Step 3: Find the contract cost.

Contract cost = actual cost + overhead fee

Contract cost = 110,000 + 7,000

Contract cost = $117,000 (this is the final value of the contract/procurement)

EXERCISE PROBLEMS

1. Your company has negotiated a cost plus incentive fee contract with a vendor. The contract has the following details:

   Target cost: $500,000

   Target fee: $60,000

   Target price: $600,000

   Share ratio: buyer 70% and vendor 30%

   Actual cost: $400,000

   What is the contract cost to your company?

   A. $490,000
   B. $500,000
   C. $560,000
   D. $600,000

2. Your company has a cost plus incentive fee contract with a vendor. The contract has a target cost of $200,000 and a target fee set at 25% of that total. If the target price is $300,000, the share ratio between buyer and vendor is set at 70/30, and the actual cost of the contract is $350,000, how much does the vendor make on this contract?

   A. $300,000
   B. $350,000
   C. $355,000
   D. $415,000

3. Your company has a cost plus incentive fee contract with a vendor in which profits are shared in the following ratio: buyer 60%,

vendor 40%. The contract has a target cost of $220,000, of which $200,000 is the estimated cost and $20,000 represents the vendor's 10% fixed fee. If the actual cost was only $160,000, what was the total cost of the project?

A. $160,000
B. $200,000
C. $204,000
D. $220,000

4. You are a vendor who has negotiated a fixed price plus incentive fee contract to do some work for a company. The contract has a target cost of $300,000 and a ceiling price has been set at $375,000. Your target fee is 50,000. The share ratio between buyer and vendor is set as follows: buyer 70% and vendor 30%.

   The actual cost on the procurement is $350,000. What is the final price the company will have to pay you to close the contract?

A. $325,000
B. $375,000
C. $385,000
D. $415,000

5. Your company has a cost plus incentive fee contract with a vendor. The target cost is $290,000; the target fee is $20,000, and the share ratio between buyer and vendor is set at 80/20. There is a maximum fee of $30,000 and a minimum fee of $5000. If the actual cost is $250,000, what fee does your company have to pay the vendor?

A. $12,000
B. $18,000
C. $20,000
D. $28,000

6. Your company has a cost plus incentive fee contract with a vendor. The target cost is $150,000, the target fee is $10,000, and the share ratio between buyer and vendor is set at 80/20. There is a maximum fee of $22,000 and a minimum fee of $10,000. If the actual cost is $100,000, what does your company owe the vendor?

A. $20,000
B. $22,000
C. $12,000
D. $10,000

7. Your company has a fixed price plus incentive fee contract with a vendor. The contract has the following details:

Target cost: $200,000

Target profit: $50,000

Target price: $230,000

Ceiling price: $270,000

Share ratio: buyer 70% and vendor 30%

Actual cost: $170,000

What is the vendor's total profit from this contract?

A. $41,000
B. $59,000
C. $9,000
D. $50,000

8. In a fixed price plus incentive fee contract, the target cost is $300,000 and the target fee is 10%. A price ceiling has been set at $330,000. The share ratio percentage for the vendor is 30%. If the actual cost is $310,000, what is the final value of this procurement?

A. $310,000
B. $330,000
C. $317,000
D. $337,000

9. In a cost plus incentive fee contract, the target cost is $100,000 and the target profit is 10%. A price ceiling has been set at $130,000. The share ratio percentage for the vendor is 20%. If the actual cost includes an overrun of $10,000, what is the final value of this procurement?

A. $118,000
B. $128,000
C. $122,000
D. $120,000

10. Your company has a cost plus incentive fee contract with a vendor. The target cost is $150,000 and the target fee is $20,000. The share ratio between buyer and vendor is set at 80/20. There is a

maximum fee of $30,000 and a minimum fee of $5000. If the actual cost is $180,000, what fee does your company have to pay the vendor?

A. $6000
B. $14,000
C. $20,000
D. $26,000

Exercise Answers

1. **A**

   Step 1: Find the incentive.

   Since the actual cost is less than the target cost,

   Incentive = (target cost − actual cost) × vendor's share ratio percentage

   Incentive = (500,000 − 400,000) × 30/100

   Incentive = 100,000 × 0.3 = $30,000

   Step 2: Find the overhead fee.

   Overhead fee = target fee + incentive

   Overhead fee = 60,000 + 30,000

   Overhead fee = $90,000

   Step 3: Find the contract cost.

   Contract cost = actual cost + overhead fee

   Contract cost = 400,000 + 90,000

   Contract cost = **$490,000**

2. **C**

   Step 1: Find the incentive.

   Because the actual cost is greater than the target cost, there is a cost overrun and thus a negative incentive.

   Incentive = (target cost − actual cost) × vendor's share ratio percentage

   Incentive = (200,000 − 350,000) × 30/100

   Incentive = −150,000 × 0.3 = − $45,000

Step 2: Find the overhead fee.

A cost overrun means that the vendor's negative incentive (from Step 1) must be deducted from the target fee or profit.

Overhead fee = target profit + incentive

Overhead fee = (25 % of target cost) + (−45,000)

Overhead fee = 50,000 − 45,000 = $5,000

Step 3: Find the contract cost.

Contract cost = actual cost + overhead fee

Contract cost = 350,000 + 5,000

Contract cost = **$355,000**

3. **C**
   Step 1: Find the incentive.

Incentive = (target cost − actual cost) × vendor's share ratio percentage

Incentive = (220,000 − 160,000) × 40/100

Incentive = 60,000 × 0.4 = $24,000

Step 2: Find the overhead fee.

Overhead fee = target fee + incentive

Overhead fee = 20,000 + 24,000

Overhead fee = $44,000

Step 3: Find the contract cost.

Contract cost = actual cost + overhead fee

Contract cost = 160,000 + 44,000

Contract cost = **$204,000**

4. **B**
   Step 1: Find the Incentive.

Because the actual cost is greater than the target cost, there is a cost overrun.

Incentive = (target cost − actual cost) × vendor's share ratio percentage

Incentive = (300,000 − 350,000) × 30/100

Incentive = −50,000 × 0.3 = −$15,000

Step 2: Find the overhead fee.

A cost overrun means that the vendor's negative incentive (from Step 1) must be deducted from the target fee or profit.

Overhead fee = target fee + incentive

Overhead fee = 50,000 + (−15,000)

Overhead fee = $35,000

Step 3: Find the contract cost.

Contract cost = actual cost + overhead fee

Contract cost = 350,000 + 35,000

Contract cost = $385,000

Step 4: Compare the ceiling price with the contract cost.

Ceiling price < contract cost

Because the ceiling price is less than the contract cost, the cost of the contract will be limited to the ceiling price, which is **$375,000**. Had there been no ceiling price or if the contract cost were less than the ceiling price, the contact cost would have been $385,000 as computed in Step 3.

5. **D**
   Step 1: Find the incentive.

   Incentive = (target cost − actual cost) × vendor's share ratio percentage

   Incentive = (290,000 − 250,000) × 20/100

   Incentive = 40,000 × 0.2 = $8000

   Step 2: Find the overhead fee.

   Overhead fee = target fee + incentive

   Overhead fee = 20,000 + 8000

   Overhead fee = $28,000

   Step 3: Compare the overhead fee with the ceiling fee (maximum fee).

   Maximum fee > overhead fee

Because the calculated overhead fee does not exceed the maximum fee specified in the contract, the amount payable to the vendor would be **$28,000** as computed in Step 2.

6. **A**

   Step 1: Find the incentive.

   Incentive = (target cost − actual cost) × vendor's share ratio percentage

   Incentive = (150,000 − 100,000) × 20/100

   Incentive = 50,000 × 0.2 = $10,000

   Step 2: Find the overhead fee.

   Overhead fee = target profit + incentive

   Overhead fee = 10,000 + 10,000

   Overhead fee = $20,000

   Step 3: Compare the overhead fee with the ceiling fee.

   Ceiling fee > overhead fee

   Because the ceiling fee is greater than the actual overhead fee, the amount the company owes the vendor would be **$20,000**, as computed in Step 2.

7. **B**

   Step 1: Find the incentive.

   Incentive = (target cost − actual cost) × vendor's share ratio percentage

   Incentive = (200,000 − 170,000) × 30/100

   Incentive = 30,000 × 0.3 = 9,000

   Step 2: Find the overhead fee due to the vendor.

   Overhead fee = target profit + incentive

   Overhead fee = 50,000 + 9,000 = **$59,000** (this is the vendor's profit from this contract)

8. **B**

   Step 1: Find the incentive.

   Because the actual cost is greater than the target cost, there is a cost overrun.

Incentive = (target cost – actual cost) × vendor's share ratio percentage

Incentive = (300,000 – 310,000) × 30/100

Incentive = –10,000 × 0.3 = –$3,000

Step 2: Find the overhead fee.

A cost overrun means that the vendor's negative incentive (from Step 1) must be deducted from the target fee or profit.

Overhead fee = target fee + incentive

Overhead fee = 30,000 + (–3,000)

Overhead fee = $27,000

Step 3: Find the contract cost.

Contract cost = actual cost + overhead fee

Contract cost = 310,000 + 27,000

Contract cost = $337,000 (this is the total cost of the procurement)

Step 4: Compare the contract cost with the price ceiling.

Contract cost > price ceiling

Because the contract cost is greater than the price ceiling of $330,000, the final value is **$330,000**.

9. **A**

Step 1: Find the incentive.

The question indicates that the actual cost includes a $10,000 overrun, which means that it is $10,000 more than the target cost. So the actual cost = $110,000.

Incentive = (target cost – actual cost) × vendor's share ratio percentage

Incentive = (100,000 – 110,000) × 20/100

Incentive = –10,000 × 0.2 = –$2,000

Step 2: Find the overhead fee.

A cost overrun means that the vendor's negative incentive (from Step 1) must be deducted from the target fee or profit.

Overhead fee = target fee + incentive

Overhead fee = 10,000 + (–2,000)

Overhead fee = $8,000

Step 3: Find the contract cost.

Contract cost = actual cost + overhead fee

Contract cost = 110,000 + 8,000

Contract cost = $118,000 (this is the total cost of the procurement)

Step 4: Compare the total cost with the price ceiling.

The total cost of **$118,000** is less than the price ceiling of $130,000, so the contract cost is as computed in Step 3.

10. **B**
   Step 1: Find the incentive.

Incentive = (target cost − actual cost) × vendor's share ratio percentage

Incentive = (150,000 − 180,000) × 20/100

Incentive = −30,000 × 0.2 = −$6,000

Step 2: Find the overhead fee.

Overhead fee = target profit + incentive

Overhead fee = 20,000 + (−6,000)

Overhead fee = $14,000

Step 3: Compare the overhead fee with the ceiling fee.

Ceiling fee > overhead fee

Since the ceiling fee is greater than the actual fee; the total fee payable to the vendor would be **$14,000**, as computed in Step 2.

2.9 POINT OF TOTAL ASSUMPTION

Main Concept

Point of Total Assumption (PTA) is a point on the cost curve of a procurement beyond which the seller (vendor/contractor) assumes all the costs of an overrun. It's basically a tipping point or breaking point, which is also referred to as the "most pessimistic cost" beyond which the seller will lose a dollar of profit (either reducing the seller's profit or increasing the seller's loss) for each additional dollar spent. The PTA is pessimistic in the sense that under reasonable circumstances, the cost should not exceed it. Once the PTA is reached, the buyer is obligated to pay only the ceiling price specified in the procurement contract.

PTA is relevant to both fixed price plus incentive fee contracts and cost reimbursable contracts. Further details along with the formulas to calculate PTA are provided below.

Determining the PTA is an important function of the Project Manager during procurement management.

Fixed Price Plus Incentive Fee Contract

A fixed price plus incentive fee (FPIF) contract is usually preferred when the scope of the project is defined and well documented. The vendor's incentive is to meet or exceed predefined performance goals such as scheduled milestone dates or quality levels. The PTA also protects the buyer from being at a disadvantage if the project costs go out of control, by charging the vendor for costs that exceed the PTA.

An FPIF contract specifies a target cost, a target profit, a target price, a ceiling price, and the share ratio for how a cost overrun or savings would be split between the buyer and vendor. The PTA is calculated as the target cost plus the difference between the ceiling and target prices, divided by the buyer's percent of the share ratio.

PTA = target cost + [(ceiling price – target price)/buyer's percent of share ratio]

Cost Reimbursable Contract

A cost reimbursable (CR) contract is usually preferred for projects in which the entire scope of the work involved in the procurement isn't known in advance. The onus is on the vendor to do all the work and the buyer will reimburse the cost incurred, in addition to a fixed fee. The PTA protects the buyer from being charged for costs that go way beyond the budget.

A CR contract specifies a ceiling price, a target cost, a fixed fee and the share ratio for a cost overrun or savings. The PTA is calculated as the target cost plus the difference between the ceiling price and the sum of the target cost and the fixed fee, divided by the buyer's percent of the share ratio.

$$PTA = \text{target cost} + \{(\text{ceiling price} - (\text{target cost} + \text{target fee}))/\text{buyer's percent of share ratio}\}$$

Point of Total Assumption Problems on the PMP Exam

TYPES OF PROBLEMS TO EXPECT

There are only two types of PTA problem on the PMP exam:

(1) Computing the PTA for a fixed price plus incentive fee (FPIF) contract
(2) Computing the PTA for a cost reimbursable (CR) contract

DIFFICULTY LEVEL

The problems in the PTA section have a moderate level of difficulty.

NUMBER OF PROBLEMS TO EXPECT

• Expect a maximum of one problem in the exam from this section.

FORMULAS

The buyer/vendor share ratio and contract ceiling price are extremely significant parameters of a contract and have the power to influence the PTA and hence the profitability of the contract.

For fixed price plus incentive fee contracts:

$$PTA = \text{target cost} + [(\text{ceiling price} - \text{target price})/\text{buyer's percent of share ratio}]$$

For cost reimbursable contracts:

$$PTA = \text{target cost} + [(\text{ceiling price} - (\text{target cost} + \text{target fee}))/\text{buyer's percent of share ratio}]$$

INSIDER TIPS

• Although at first glance it appears that the formulas for PTA for CR and FPIF contracts are different, they are essentially similar. The target price in an FPIF contract is equivalent to the target cost in a CR contract; and FPIF contracts do not have a target fee. So effectively,

PTA (FPIF) = PTA (CR) (where target fee = 0)

You could just memorize the formula for a CR contract and substitute the target fee with a 0.

TIP: *Be aware of the fact that the share ratio always reflects the buyer's share first and then the vendor's share of a cost overrun or savings.*

SAMPLE SOLVED PROBLEMS

1. In a fixed price plus incentive contract, the target cost is $300,000 with a target profit of 10%. The target price is $330,000 and there is a ceiling price of $350,000. The buyer's share of cost overruns and savings is 80%. What is the Point of Total Assumption of this procurement?

Step 1: Find the PTA.

In this case, all you need to do is to apply the formula for a fixed price plus incentive fee contract.

PTA = target cost + [(ceiling price – target price)/buyer's percent of share ratio]

PTA = 300,000 + [(350,000 – 330,000)/0.80]

PTA = 300,000 + (20,000/0.80)

PTA = $325,000

NOTE: *The target profit of 10% given in the problem is not required to calculate the PTA. Keep an eye out for questions that provide more data than you need to compute the answer.*

2. Your company has negotiated a cost reimbursable contract with a vendor with a target cost of $300,000 and a target fee of 10%. The maximum price is set at $350,000 and the buyer's negotiated share ratio is 80%. What is the Point of Total Assumption of this procurement?

Step 1: Determine the target fee.

The target fee is given as 10%, which is 10% of the target cost.

So target fee = 10% of $300,000 = $30,000

Step 2: Find the PTA.

In this case, all you need to do is to apply the formula for a cost reimbursable contract.

PTA = target cost + [(ceiling price - (target cost + target fee))/buyer's percent of share ratio]

PTA = 300,000 + [(350,000 − (300,000 + 30,000))/0.80]

PTA = $325,000

EXERCISE PROBLEMS

1. In a fixed price plus incentive fee contract, the target cost is $100,000. There is a target profit of 10%, a target price of $110,000, and a ceiling price of $120,000. The share ratio percentage for the buyer is 80%. What is the point of total assumption of this procurement?

   A. $108,000
   B. $100,000
   C. $112,500
   D. $120,000

2. Consider a fixed price plus incentive fee contract with the following details:

   Target cost: $800,000

   Target price: $640,000

   Share ratio: buyer 60%, vendor 40%

   Ceiling price: $940,000

   What is the Point of Total Assumption?

   A. $1,000,000
   B. $940,000
   C. $900,000
   D. $1,300,000

3. Your company has negotiated a cost reimbursable contract with a target cost of $600,000 and a target fee of $20,000. The buyer and vendor split the cost of overruns and any potential savings. If the ceiling price for this procurement is set at $700,000 what is the Point of Total Assumption?

A. $760,000
B. $600,000
C. $700,000
D. $620,000

4. Given the following information for a cost reimbursable contract, what would be the break point after which the seller would assume all cost overruns?

Target cost: $700,000

Target fee: $50,000

Share ratio: buyer 60%, seller 40%

Ceiling price: $900,000

A. $900,000
B. $950,000
C. $800,000
D. $850,000

5. In a fixed price plus incentive fee contract, the target cost is $500,000. There is a target price of $560,000 and a ceiling price of $580,000. The share ratio percentage for the buyer is 80%. What is the Point of Total Assumption of this procurement?

A. $500,000
B. $580,000
C. $520,000
D. $525,000

6. A fixed price incentive fee contract is targeted to cost $400,000. The price of the contract is targeted to be $440,000 with a maximum of $500,000. The buyer's share of savings and cost overruns is 60%. What is the Point of Total Assumption of this procurement?

A. $500,000
B. $440,000
C. $400,000
D. $520,000

7. Your company has reached consensus on a cost reimbursable contract with one of its vendors. The target cost for this procurement is $700,000 with a target price of $780,000, a target fee of 10%,

and a price ceiling of $900,000. The share ratio is 50% for the buyer. What is the Point of Total Assumption of this procurement?

A. $700,000
B. $770,000
C. $960,000
D. $880,000

8. For a cost reimbursable contract, the target cost is $300,000, the target price is $350,000, the target fee is 15% and the ceiling price is $400,000. The share ratio is 50% for the buyer. What is the point at which the seller would assume responsibility for all cost overruns?

A. $500,000
B. $410,000
C. $300,000
D. $600,000

9. Given the following data about a cost reimbursable contract, what would be the Point of Total Assumption?

Target cost: $600,000
Target fee: $60,000
Share ratio: buyer 80%, seller 20%
Ceiling price: $720,000

A. $600,000
B. $740,000
C. $660,000
D. $675,000

10. What would be the Point of Total Assumption for a fixed price incentive fee contract with the following details?

Target cost: $400,000
Target price: $450,000
Share ratio: buyer 70%, seller 30%
Ceiling price: $520,000

A. $500,000
B. $460,000
C. $300,000
D. $600,000

Exercise Answers

1. **C**

   Use the PTA formula for an FPIF contract:

   PTA = target cost + [(ceiling price – target price)/buyer's percent of share ratio]

   PTA = 100,000 + [(120,000 – 110,000)/0.8]

   PTA = **$112,500**

2. **D**

   Use the PTA formula for an FPIF contract:

   PTA = target cost + [(ceiling price – target price)/buyer's percent of share ratio]

   PTA = 800,000 + [(940,000 – 640,000)/0.6]

   PTA = **$1,300,000**

3. **A**

   Use the PTA formula for a CR contract:

   PTA = target cost + [(ceiling price – (target cost + target fee))/buyer's percent of share ratio]

   PTA = 600,000 + [(700,000 – (600,000 + 20,000)/0.5]

   PTA = **$760,000**

4. **B**

   Use the PTA formula for a CR contract:

   PTA = target cost + [(ceiling price – (target cost + target fee))/buyer's percent of share ratio]

   PTA = 700,000 + [(900,000 – (700,000 + 50,000))/0.6]

   PTA = **$950,000**

5. **D**

   Use the PTA formula for an FPIF contract:

   PTA = target cost + [(ceiling price – target price)/buyer's percent of share ratio]

   PTA = 500,000 + [(580,000 – 560,000)/0.8]

   PTA = **$525,000**

6. **A**

   Use the PTA formula for an FPIF contract:

PTA = target cost + [(ceiling price – target price)/buyer's percent of share ratio]

PTA = 400,000 + [(500,000 – 440,000)/0.6]

PTA = **$500,000**

7. **C**

For a CR contract you need to account for the target fee as well. Use the PTA formula:

PTA = target cost + [(ceiling price – (target cost + target fee))/buyer's percent of share ratio]

PTA = 700,000 + [(900,000 – (700,000 + 70,000))/0.5]

PTA = **$960,000**

NOTE: *The target price of $780,000 given in the problem is not required to calculate the PTA. Keep an eye out for questions that provide more data than you need to compute the answer.*

8. **B**

Because this is a cost reimbursable contract, you need to account for the target fee as well. Use the PTA formula:

PTA = target cost + [(ceiling price – (target cost + target fee))/buyer's percent of share ratio]

PTA = 300,000 + [(400,000 – (300,000 + 45,000)) /0.5]

PTA = **$410,000**

9. **D**

Use the PTA formula for a CR contract:

PTA = target cost + [(ceiling price – (target cost + target fee))/buyer's percent of share ratio]

PTA = 600,000 + [(720,000 – (600,000 + 60,000))/0.8]

PTA = **$675,000**

10. **A**

Use the PTA formula for an FPIF contract:

PTA = target cost + [(ceiling price – target price)/buyer's percent of share ratio]

PTA = 400,000 + [(520,000 – 450,000)/0.7)]

PTA= **$500,000**

2.10 DETERMINE COMMUNICATION CHANNELS

Main Concept

Almost 90% of a Project Manager's time is spent communicating, so it is vital to be aware of the nature and the number of communication channels that are available. The number of channels correlates directly to the number of individuals on the team. The number of channels can be determined by counting all the different combinations of communications that can take place between team members.

Figure 2.9 depicts the communication channels that exist in a team of size 2, 3, 4, and 5 members, respectively. The bigger the team, the greater the number of channels and the more challenging it becomes for the Project Manager to communicate effectively.

TIP: *Always be aware of the number of communication channels in order to manage project communications effectively.*

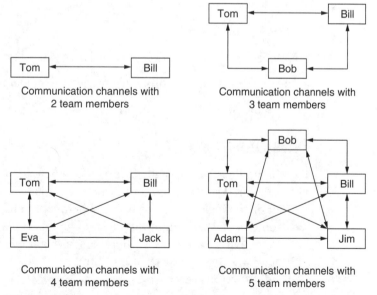

Communication channels with
2 team members

Communication channels with
3 team members

Communication channels with
4 team members

Communication channels with
5 team members

Figure 2.9 Comparing communication channels with 2, 3, 4, and 5 team members.

Communication Channels Problems on the PMP Exam

TYPES OF PROBLEMS TO EXPECT

There are three types of communication channels problems that you can expect to encounter on the PMP exam:

(1) Computing the total number of communication channels
(2) Computing the increase in the number of communication channels
(3) Computing the decrease in the number of communication channels

DIFFICULTY LEVEL

The problems on communication channels will be straightforward. This topic is arguably one of the easiest and least math-intensive of all topics on the PMP exam.

NUMBER OF PROBLEMS TO EXPECT

Expect one or two questions regarding communication channels on the PMP exam. Chances are you will be asked to compute the increase in the number of communication channels.

FORMULAS

Number of communication channels = $[N (N - 1)]/2$

where N is the number of people in the project team.

The same formula can be written as $(N^2 - N)/2$

INSIDER TIPS

- On the PMP exam, make sure you compute the required value. Often the exam question will require you to compute the increase in the number of channels instead of the total number of channels.
- Remember that communication channels grow *exponentially* and not linearly when new members are added. So when you add members, the increase in the number of communication channels is dramatic.

SAMPLE SOLVED PROBLEMS

1. You are a Project Manager for a construction project. Two contractors report to you and your team also includes 2 other engineers. How many communication channels do you have to manage?

This is a simple question requiring you to first determine the team size and then compute the number of communication channels for that team.

The team includes the Project Manager, the two contractors, and the two engineers. So it is a team of 5 (including yourself).

Number of communication channels = $[N(N-1)]/2$

where N is the number of people in the team.

Thus the number of channels = $[5(5-1)]/2 = 10$ channels

2. The Project Manager of a large IT initiative has a team of 12, including 6 developers and 2 quality assurance professionals. A recent quality audit suggests that additional resources are needed to complete quality inspections and to help the project stay on schedule, albeit at a higher cost. How many more channels would the Project Manager have to manage given the addition of 2 new quality professionals?

Note that the question asks how many *more* channels would need to be managed. This means that you are looking for the number of new channels added as a result of the addition of the two new members to the team. The solution involves (a) computing the number of channels for the original team of 12 members, (b) computing the number of channels for the new team of 14 members, and (c) finding the difference.

Step 1: Find the number of channels for the original team of 12.

Total number of communication channels is $[N(N-1)]/2$

where N is the number of people in the team.

Originally $N = 12$

Thus the number of channels = $[12(12-1)]/2 = 66$.

Step 2: Find the number of channels for the new team of 14.

Total number of communication channels is $[N(N-1)]/2$

where N is the number of people in the team.

With the new additions to the team, $N = 14$

Thus the number of channels is now $[14(14 − 1)]/2 = 91$.

Step 3: Find how many channels were added.

The number of channels added to the project is the difference between the results in Step 1 and Step 2.

Increased channels = number of channels in new team − original number of channels

Increased Channels = $91 − 66 = 25$

So 25 new channels will be created as a result of adding 2 new individuals to the project.

3. You are the Project Manager for a team of 8 on a software development project that is in the closing stage. As a result of the project coming to an end, 2 of the individuals on the team are being released. How many fewer communication channels will the manager have to manage?

Note that the question asks how many *fewer* channels will need to be managed. So your goal is to compute the *difference* in the communication channels resulting from the release of the 2 individuals. The solution involves (a) computing the number of channels for the original team of 8, (b) computing the number of channels after the release of 2 individuals, and (c) finding the difference.

Step 1: Find the number of channels for the original team of 8.

Total number of communication channels is $[N (N − 1)]/2$

where N is the number of people in the team.

Originally $N = 8$

Thus the number of channels = $[8(8 − 1)]/2 = 28$.

Step 2: Find the number of channels for the new team of 6 (after 2 individuals are released).

Total number of communication channels is $[N (N − 1)]/2$

where N is the number of people in the team.

After 2 individuals are released, $N = 6$

Thus the number of channels = $[6(6 − 1)]/2 = 15$.

Step 3: Find how many fewer channels the manager will need to manage.

The number by which the channel total is reduced is the difference between the results in Step 1 and Step 2.

Reduction in the number of Channels = Original number of channels – number of channels after individuals left = 28 – 15 = 13.

So after two individuals leave the team, the manager will need to manage 13 fewer communication channels.

EXERCISE PROBLEMS

1. A Project Manager has 20 members on his team, of whom 5 are located offsite. Based on recommendations from senior management, he/she decides to add 2 more members to his testing team. How many additional communication channels are introduced as a result of this organizational change?

   A. 126
   B. 41
   C. 231
   D. 190

2. A Project Manager has 4 engineers, 2 quality analysts, and 2 senior managers in her team. There are also 2 external stakeholders to whom he/she must communicate the project status. Including herself as a team member, how many communication channels should he she plan for?

   A. 90
   B. 45
   C. 55
   D. 0

3. As a proactive Project Manager, you decide to create a communications management plan to ensure timely distribution of information. You are the leader of a team of 8. How many channels should you plan for?

   A. 28
   B. 56
   C. 72
   D. 64

4. After adopting the Agile methodology, your team of 8 characters to Richmond, Virginia. You also relocate 2 new team members from Cleveland to Richmond. How many more communication channels should you plan for?

   A. 28
   B. 45
   C. 73
   D. 17

5. Your have a team of 8 in Fairfax, Virginia. After corporate restructuring, 2 of your team members are assigned to a different team. How many communication channels should you plan for now?

   A. 15
   B. 28
   C. 1
   D. 17

6. How many communication channels should you plan for a team of 7?

   A. 42
   B. 28
   C. 21
   D. 17

7. You are newly promoted as a Project Manager within your organization. Your mentor coaches you on the importance of communication and asks you to compute the number of communication channels for your project. What mathematical formula will you use to ensure you always get the right answer?

   A. $[N(N+1)]/2$
   B. $N \times N/2$
   C. $[N(N-1)]/2$
   D. $(N-1)/2$

8. A Project Manager has 10 members on his team. After adding 3 more contractors to the team, how many total communication channels does he need to plan for?

   A. 156
   B. 78
   C. 45
   D. 33

9. A Project Manager has 10 members on her team. After adding 3 more contractors to the team, how many more communication channels does she need to plan for?

    A. 33
    B. 78
    C. 45
    D. 156

10. A Project Manager has 10 members on his team. After down sizing the team by 4, how many communication channels does he need to plan for?

    A. 45
    B. 14
    C. 91
    D. 15

Exercise Answers

1. **B**

    The total number of team members is now 22. The 5 who are off-site are included among the original 20 and their location is just an extra piece of information. The original number of communication channels was $20 \times (20 - 1)/2 = 20 \times 19/2 = 190$. The new number of communication channels is $22 \times (22 - 1)/2 = 22 \times 21/2 = 231$. The number of channels added is $231 - 190 = $ **41.**

2. **C**

    The total number of people in the PM's team is $4 + 2 + 2 + 2 + 1 = 11$ (don't forget the PM). So $11 \times (11 - 1)/2 = $ **55.**

3. **A**

    This problem is straightforward. All you need to do is substitute values in $N(N - 1)/2$. So we get $(8 \times 7)/2 = $ **28.**

4. **D**

    The total number of team members was originally 8. The number of communication channels was $8 \times (8 - 1)/2 = 28$. After 2 additional members join the team, the new number of communication channels is $10 \times (1 - 1)/2 = 45$. The number of channels added is $45 - 28 = $ **17.**

5. **A**

   The total number of remaining team members left is $8 - 2 = 6$. The number of communication channels is $6 \times (6 - 1)/2 = \mathbf{15}$.

6. **C**

   The number of communication channels is $7 \times (7 - 1)/2 = \mathbf{21}$.

7. **C**

   Answer choice C is the correct formula.

8. **B**

   The total number of team members is now $10 + 3 = 13$. The number of communication channels is $13 \times (13 - 1)/2 = \mathbf{78}$.

9. **A**

   The total number of team members was originally 10. The number of communication channels was $10 \times (10 - 1)/2 = 45$. After 3 additional members join the team, the new number of communication channels is $13 \times (13 - 1)/2 = 78$. The number of channels added is $78 - 45 = \mathbf{33}$.

10. **D**

    The total number of remaining team members is $10 - 4 = 6$. The number of communication channels is $6 \times (6 - 1)/2 = \mathbf{15}$.

CHAPTER 3

Execution

This chapter covers	1. Earned value analysis: assessment
	• Cost variance
	• Schedule variance
	• Cost performance index
	• Schedule performance index
	2. Earned value analysis: forecast
	• Estimate at completion
	• Estimate to complete
	• Variance at completion

INTRODUCTION

The prerequisite for a successful project is the creation of a realistic, "bought-into," and approved Project Management Plan. But the plan itself does not guarantee success; it is the execution of the plan that determines project success or failure. The process of execution involves performing the tasks outlined in the Project Management Plan with the goal of fulfilling the project's objectives.

In this phase, the Project Manager plays the critical role of coordinator and facilitator, directing various activities, most of which might be technical in nature. The execution phase results in the creation of the product of the project.

In the following sections, we will introduce some key mathematical concepts that are relevant to the execution process and the task of assessing or forecasting a project's financial health. The sections below discuss Earned Value Analysis that can be used for both assessing and forecasting several key attributes of a project.

Earned Value Analysis: Assessment

Earned Value Analysis is a tool that provides insight into a project's true value and financial status. The technique blends together performance data relevant to project scope, schedule, and budget (cost) to present unified objective performance parameters. It is usually used to assess project status and performance on a particular date, but there are parameters that can also be used to forecast project performance. The sections below discuss Earned Value Analysis from the perspective of assessing current and past project performance and also from the perspective of forecasting future performance.

Earned Value Analysis provides the Project Manager with a preview into the current state of the project. There are four aspects to this analysis: Cost Variance, Schedule Variance, Cost Performance Index, and Schedule Performance Index. Cost Variance enables the Project Manager to determine whether the project is over budget, on budget, or under budget. Schedule Variance shows whether the project is behind schedule, on schedule, or ahead of schedule. Cost Performance Index provides information on the efficiency of the amount spent and the value recovered. Schedule Performance Index shows the rate at which the project is progressing.

Earned Value Analysis: Forecast

Future Earned Value Analysis aims to predict the performance of a project based on the current Earned Value Analysis. There are three aspects to this analysis. Estimate at Completion enables the Project Manager to estimate how much the project will cost based on the current project history. Estimate to Complete tracks the amount the project will cost. Variance at Completion analyzes the deviation from the cost baseline at the end of the project.

3.1 EARNED VALUE ANALYSIS: ASSESSMENT

Main Concept

As described above, Earned Value Analysis provides a way to assess a project's performance by examining parameters that combine the effect of that project's scope, schedule, and cost on performance. These parameters can be compared to established baselines to assess the health of the project. The assessment part of this analysis focuses strictly on determining project performance status on a given date (current or past). There is a mathematical approach to this technique involving evaluation of variances and performance indices from a cost and schedule perspective. Before we diagnose the cost and schedule variance or performance indices, it is vital to understand certain key underlying concepts, as follows.

Planned Value

Planned Value (PV) represents the estimated value of work that must be done on the project. It represents the estimated Planned Value for all the work that is scheduled on the project. It is determined during the early stages of the project; prior to the work actually being accomplished and can also serve as a baseline. For example, suppose that the work to be done on a project to build a web site is estimated to be valued at $9,000. This estimated value is the PV, which may also be referred to as the budgeted cost of work scheduled (BCWS).

Earned Value

The Earned Value (EV) of a project represents the value of work actually accomplished at a given time. Suppose again that a project has a Planned Value of $9,000. If at the end of the first month only one-third of the work is accomplished, the EV at the end of that first month would be $3,000 (1/3 × $9,000).

EV may also be referred to as the budgeted cost of work performed (BCWP).

Actual Cost

Actual Cost (AC) represents the amount of money actually spent on a project to date.

AC may also be referred to as the actual cost of work performed (ACWP).

Cost Variance

Cost Variance (CV) is the difference between the value of work actually accomplished (EV) and the Actual Cost (AC) of a project. A positive Cost Variance implies that the project is under budget, while a negative Cost Variance implies that the project has overshot its budget. When the Cost Variance is zero, the project is right on budget.

$$\text{Cost Variance (CV)} = \text{Earned Value (EV)} - \text{Actual Cost (AC)}$$

Cost Performance Index

Cost Performance Index (CPI) analyzes the cost efficiency of a project. It is represented by the ratio of the Earned Value to the Actual Cost of the project and depicts the value you are getting out of every dollar that is spent on the project. A CPI greater than 1 implies that you are getting more value for work performed than the cost incurred in doing that work and so are more likely to be under budget for the value delivered in the project. A CPI less than 1 implies that the value of the work accomplished is less than the cost incurred, which means that you are burning cash at a faster rate than you are earning value and are hence more likely to overshoot your budget. A CPI equal to 1 implies that you are earning a dollar for every dollar that is being spent on the project and are most likely to be right on budget (See Fig. 3.1). A word of caution when using CPI to determine cost efficiency: Flawed estimates and budgets can contribute to a misleading CPI.

$$\text{Cost Performance Index (CPI)} = \text{Earned Value (EV)} \div \text{Actual Cost (AC)}$$

Schedule Variance

Schedule Variance (SV) is the difference between the value of work actually accomplished and the estimated value of work that was planned. A positive Schedule Variance implies that the project is ahead of schedule, while a negative Schedule Variance implies that the project is behind schedule, in terms of cost (See Fig. 3.2). When the Schedule Variance is zero, the project is right on schedule.

$$\text{Schedule Variance (SV)} = \text{Earned Value (EV)} - \text{Planned Value (PV)}$$

Even though SV refers to the schedule, it represents how well the project is doing in terms of the cost of the work scheduled. So in other words, SV indicates whether or not the cost incurred on the project is over the scheduled cost.

Schedule Performance Index

Schedule Performance Index (SPI) analyzes the schedule efficiency of a project in terms of its cost. It is the ratio of the Earned Value to the Planned

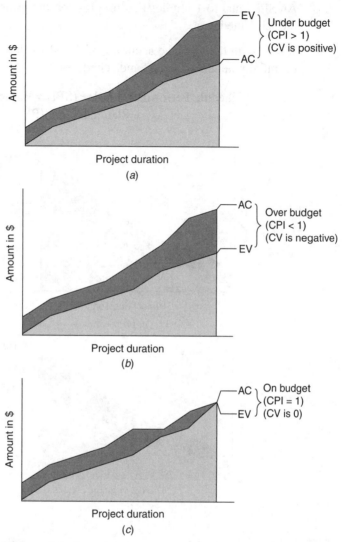

Figure 3.1 Cost Variance: (*a*) Positive CV, under budget; (*b*) negative CV, over budget; (*c*) zero CV, on budget.

Value of the project and depicts the actual progress made on the project with respect to the planned progress, i.e., how much of the project work has been completed over the scheduled cost. SPI is an efficiency index that shows the efficiency of the time utilized on the project. An SPI greater than 1 implies that the rate of progress is better than what was planned; which means the project is ahead of schedule. An SPI less than 1 implies that the progress is slower than originally planned, which means that you are behind schedule.

An SPI equal to 1 implies that progress on the project is exactly in line with the scheduled cost.

Just as with the CPI, the accuracy of the SPI value depends on the accuracy of initial planned estimates and schedules.

$$\text{Schedule Performance Index (SPI)} = \text{Earned Value (EV)} \div \text{Planned Value (PV)}$$

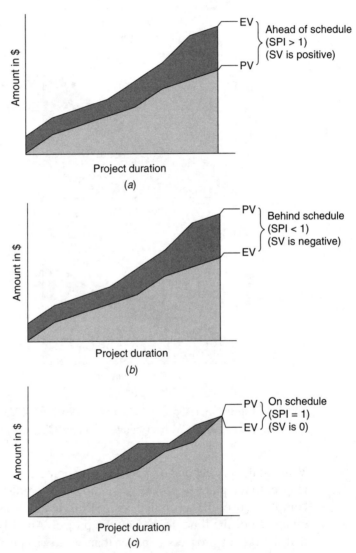

Figure 3.2 Schedule Variance: (*a*) Positive SV, ahead of schedule; (*b*) negative SV, behind schedule; (*c*) zero SV, on schedule.

Earned Value Analysis: Assessment Problems on the PMP Exam

TYPES OF PROBLEMS TO EXPECT

On the PMP exam, you can expect four types of problems based on the concepts in this chapter:

(1) Computing the Cost Variance
(2) Computing the Schedule Variance
(3) Computing the Cost Performance Index
(4) Computing the Schedule Performance Index

DIFFICULTY LEVEL

The problems in this section are based purely on the identification and application of the right formula. So memorize the formulas and understand them. Doing so should make these problems easy to deal with.

NUMBER OF PROBLEMS TO EXPECT

Expect anywhere from two to five problems on Earned Value Analysis on the PMP exam.

FORMULAS

Earned Value Parameter	Formula
Cost Variance	$EV - AC$
Cost Performance Index	$EV \div AC$
Schedule Variance	$EV - PV$
Schedule Performance Index	$EV \div PV$

INSIDER TIPS

(1) Negative variance (cost or schedule) is always considered to be an unfavorable condition, while positive or zero variance is favorable.
(2) Indices (CPI and SPI) that are equal to or greater than 1 are favorable, while values less than 1 are unfavorable.

SAMPLE SOLVED PROBLEMS

1. You are the Project Manager for a construction project budgeted at $250,000. As of today, the project should be 30% complete, but after reviewing the status of the scheduled tasks involved in the project, it is evident that only 20% of the work has been completed. The team has spent $125,000 thus far. What is the current status of this project?

   A. It is running late, but on budget.

   B. It is on schedule and on budget.

   C. It is on schedule, but over budget.

   D. It is late and over budget.

To determine if the project is on schedule and on budget, you need to find the CPI and SPI. Before you do that though, you need to calculate the EV and PV with the given data.

Step 1: Determine EV.

The problem indicates that only 20% of the work has actually been completed. As a result only 20% of the value has been earned.

$$EV = 20\% \text{ of Budget} = 20\% \times 250{,}000 = \$50{,}000$$

Step 2: Determine PV.

The problem indicates that the Project Manager had planned to have 30% of the work completed thus far.

$$PV = 30\% \text{ of Budget} = 30\% \times 250{,}000 = \$75{,}000$$

Step 3: Determine CPI.

$$CPI = EV \div AC = 50{,}000 \div 125{,}000 = 0.4$$

CPI < 1. So the project is well over budgeted cost.

Step 4: Determine SPI.

$$SPI = EV \div PV = 50{,}000 \div 75{,}000 = 0.67$$

SPI < 1. So the project is also well behind schedule.

As a result, the correct answer is choice D: The project is late and over budget.

2. A project you are managing has a Cost Variance of $300. If you have spent $1,000 on the project and had planned to spend $1,200, what is the Schedule Variance?

A. − $400

B. $400

C. − $100

D. $100

To calculate Schedule Variance, you need both the PV and EV. Because you already have the PV, you need to now determine the EV using the given Cost Variance.

Step 1: Determine EV.

$$\text{Cost Variance} = \text{EV} - \text{AC}$$

$$\text{EV} = \text{CV} + \text{AC} = 300 + 1{,}000 = \$1{,}300$$

Step 2: Determine SV.

$$\text{Schedule Variance (SV)} = \text{EV} - \text{PV}$$

$$\text{SV} = 1{,}300 - 1{,}200 = \$100 \text{ (answer choice D)}$$

EXERCISE PROBLEMS

1. You are managing a project with a PV of $40,000, an EV of $30,000, and AC of $50,000. What is the Cost Variance for the project?

A. (−) $10,000

B. (+) $10,000

C. (+) $20,000

D. (−) $20,000

2. You are managing a project with a PV of $60,000, an EV of $70,000, and an AC of $100,000. What is the Schedule Variance for the project?

A. (−) $10,000

B. (+) $10,000

C. (+) $30,000

D. (−) $30,000

3. The PV for a project is $15,000, the EV is $30,000, and the AC is $20,000. Given this data, what is the Schedule Performance Index?

    A. 2
    B. 0.5
    C. 1.5
    D. 1

4. For a project with a PV of $5,000, an EV of $30,000, and an AC of $60,000, what is the Cost Performance Index?

    A. 0.6
    B. 0.5
    C. 2
    D. 1.2

5. You are taking over as the Project Manager for a software development project that is already underway. In talking to the team and looking at the status reports, you determine that the CPI for the project is 0.9 and the Earned Value thus far is $225,000. How much money has been spent thus far on the project?

    A. $225,000
    B. $250,000
    C. $250,500
    D. It cannot be determined.

6. What does a CPI of 0.63 on a project imply?

    A. The project costs 63% of the initially planned budget.
    B. The project is earning only $0.63 value of work for every $1 spent on the project.
    C. The project is progressing at 63% of the rate originally planned for schedule.
    D. The project is over budget by 63%.

7. You are managing a project to build a new web site for a client. At this time, the SPI for the project has been determined to be 0.4. If the PV is $1,000 and the AC is $500, what is the CPI of the project?

    A. 0.4
    B. 0.5
    C. 0.8
    D. 1

8. A project status report indicates that the Schedule Variance is (−) $20,000. If the Actual Cost is $70,000 and the Planned Value is $80,000, what is the Cost Variance?

A. (+) $10,000
B. (−) $10,000
C. (+) $25,000
D. (−) $25,000

9. You are managing a project with an SPI of 1.1. What is the best way to describe this project?

A. The project costs 110% of the initially planned budget.
B. The project is earning only $1.10 worth of value for every $1 spent on the project.
C. The project is progressing at 110% of the rate originally planned for the schedule.
D. The project is over budget by 110%.

10. A project status report indicates that the SPI of the project is 0.6. If the Planned Value is $12,000, what is the Schedule Variance?

A. (−) $4,800
B. (+) $4,800
C. (−) $12,000
D. (−) $7,200

Exercise Answers

1. D

Don't get thrown off by the PV. This is an extra piece of information that is not required to calculate the Cost Variance.

$$CV = EV - AC = 30{,}000 - 50{,}000 = (-)\ \$20{,}000$$

(A negative Cost Variance indicates that the project is over budget.)

2. B

$$SV = EV - PV$$

$$SV = 70{,}000 - 60{,}000 = \$10{,}000$$

3. A

Schedule Performance Index (SPI) $= EV \div PV$

$$SPI = 30{,}000 \div 15{,}000 = 2$$

A SPI greater than 1 implies that the project is ahead of schedule.

4. B

$$CPI = EV \div AC$$
$$CPI = 30,000 \div 60,000 = \mathbf{0.5}$$

A CPI less than 1 implies that the project is over the budget initially planned.

5. B

In this case you are provided with the CPI and the EV. So you use the formula given below to calculate the Actual Cost (AC).

$$CPI = EV \div AC$$
$$AC = EV \div CPI = 225,000 \div 0.9 = \mathbf{\$250,000}$$

$250,000 has been spent on the project thus far.

6. B

Cost Performance Index analyzes the efficiency of the funds allocated to the project. In this case, the project realizes only **$0.63** for every dollar spent on the project.

7. C

$$SPI = EV \div PV$$
$$\text{Hence } EV = SPI \times PV$$
$$\text{Earned Value} = 0.4 \times 1,000 = 400$$
$$CPI = EV \div AC$$
$$CPI = 400 \div 500 = \mathbf{0.8}$$

8. B

$$SV = EV - PV$$
$$\text{Hence } EV = SV + PV$$
$$\text{Earned Value} = (-)\,20,000 + 80,000 = \$60,000$$
$$CV = EV - AC$$
$$CV = 60,000 - 70,000 = (-)\,\mathbf{\$10,000}$$

9. C

Schedule Variance (SV) tracks the rate at which the project schedule is progressing. The project is progressing at **110%** of the rate originally planned.

10. A

$$SPI = EV \div PV$$

So,
$$EV = SPI \times PV$$
$$EV = 0.6 \times 12,000 = 7,200$$
$$SV = EV - PV$$
$$SV = 7,200 - 12,000 = (-) \text{ } \$4,800$$

3.2 EARNED VALUE ANALYSIS: FORECAST

Main Concept

Forecasting using Earned Value Analysis is a powerful tool for management to determine the feasibility of a project and to make decisions about how the project should continue (or if it should continue). The technique helps you make predictions about future performance based on current performance data. The specific parameters that can predict future performance are "Budget at Completion," "Estimate to Complete," "Estimate at Completion," and "Variance at Completion." These are discussed in greater detail below.

Budget at Completion

Budget at Completion (BAC) represents the total budget that was initially established for the project or activity. Since this budget value is set at the initial planning stages of the project, it is also referred to as the total Planned Value for the project. If a project is broken up into phases, the BAC is the sum of the budgets of each phase. The BAC is also used to determine some other parameters, such as the Estimate at Completion (EAC) discussed below.

Estimate to Complete

Estimate to Complete (ETC) is the expected cost of the work that remains to be done on the project. ETC is a parameter that should be measured as often as possible to make sure that the expected costs are in line with the established budget. The method used to calculate the ETC varies depending on the nature of the project estimates and variances.

a. **Original estimates are flawed**   When the original estimates of a project are flawed, the ETC is best calculated by analyzing the situation and making new estimates. This approach excludes the use of any mathematical formula and involves an independent analysis of all the work remaining on the project, bearing in mind the project's past project performance analysis.

b. **Project has atypical variances**   Atypical variances imply that the project had an unexpected risk or opportunity that skewed the project performance results. If the risk or opportunity is not likely to be repeated, the variance is considered to be atypical. In this case ETC is the difference between the Budget at Completion and the Earned Value.

$$ETC = BAC - EV$$

c. **Project has typical variances**   Typical variances imply that the variances that have occurred in the project are likely to be repeated in the future. In this case, ETC is the difference between the Budget at Completion and Earned Value divided by the CPI.

$$ETC = (BAC - EV) \div CPI$$

Estimate at Completion

Estimate at Completion (EAC) is the expected cost of an activity or project at completion. The EAC at any given point in time is calculated using the actual cost of the project thus far and the estimated cost to complete the remaining work. But it can also be calculated by some other methods depending on the nature of the project. The different ways to determine the Estimate at Completion are as follows:

a. **Project cost has no deviations from the budget or the rate of spending is the same**   Since the amount being spent on the project is in line with established budget, the BAC is used to compute the EAC.

$$EAC = BAC \div CPI$$

b. **Original estimates are flawed**   If original estimates of the cost of the project are flawed, it is best to make a new estimate of the cost of the remaining work in order to compute the EAC. This new ETC includes a manual analysis of all the remaining work in the project bearing in mind past performance.

$$EAC = AC + ETC$$

c. **Project has atypical variances**   Atypical variances imply that the project had an unexpected risk or opportunity that skewed the project performance results. If the risk or opportunity is not likely to be repeated, the variance is considered to be atypical. In this case EAC is

$$EAC = AC + (BAC - EV)$$

d. **Project has typical variances**   Typical variances imply that the variances that have occurred in the project are likely to be repeated in the future. In this case EAC is

$$EAC = AC + [(BAC - EV) \div CPI]$$

Variance at Completion

Variance at Completion (VAC) is an indicator of whether the project is over or under the Budget at Completion. It is the difference between the Budget at Completion and the Estimate at Completion.

$$VAC = BAC - EAC$$

If VAC > 0 the project is under budget.
If VAC < 0 the project is over budget.
If VAC = 0 the project is on budget.

Earned Value Analysis: Forecast Problems on the PMP Exam

TYPES OF PROBLEMS TO EXPECT

On the PMP exam, you can expect to encounter three types of problems based on the concepts in this section:

(1) Computing the EAC
(2) Computing the ETC
(3) Computing the VAC

DIFFICULTY LEVEL

These problems are moderately difficult. Having to remember multiple formulas and to choose the correct formula to use can be confusing.

NUMBER OF PROBLEMS TO EXPECT

Expect a maximum of two problems on this topic on the PMP exam.

FORMULAS

Formula	When to use the formula
$ETC = (BAC - EV)$	Project is based on atypical variances
$ETC = (BAC - EV) \div CPI$	Project is based on typical variances
$EAC = BAC \div CPI$	Project has no deviations from the current project
$EAC = AC + ETC$	Project is based on new estimates
$EAC = AC + (BAC - EV)$	Project is based on atypical variances
$EAC = AC + [(BAC - EV) \div CPI]$	Project is based on typical variances
$VAC = BAC - EAC$	For all project scenarios

INSIDER TIPS

• Look for the words *typical variances* and *atypical variances* when asked to compute ETC and EAC.

SAMPLE SOLVED PROBLEMS

1. As the manager of a project for a key client, you need to give your firm's senior partner an update including a forecast of the amount estimated for the remaining work. The project is budgeted for $200,000 and the CPI is 1.25. The team has burned through 20% of the budget. What is the Estimate to Complete?

   A. $105,000

   B. $120,000

   C. $145,000

   D. $180,000

The cost estimated for the work remaining on the project is the ETC. The correct formula to use is: $ETC = EAC - AC$ (because $EAC = ETC + AC$ and you are given the AC).

Step 1: Find the AC.

The actual cost is 20% of the budget.

$$AC = 20\% \times 200,000 = \$40,000$$

Step 2: Find the EAC.

$$EAC = BAC \div CPI = 200,000 \div 1.25 = \$160,000$$

Step 3: Calculate the ETC.

$$ETC = EAC - AC = 160,000 - 40,000 = \$120,000$$

2. A project has suffered a cost and schedule setback. The variances identified in the project have been found to be atypical. Given that the BAC is $6,000, the Earned Value is $4,500, and the CPI is 1.5, what is the Estimate at Completion?

Step 1: Find the AC.

The correct formula to use is

$$CPI = EV/AC$$

$$AC = EV/CPI = 4,500/1.5 = \$3,000$$

Step 2: Find the EAC.

Because the variances are atypical in nature, the correct formula to use is

$$EAC = AC + (BAC - EV) = 3,000 + (6,000 - 4,500)$$

$$EAC = \$4,500$$

EXERCISE PROBLEMS

1. A project had suffered a cost and schedule slippage. Research indicates that the variances on the project thus far are atypical in nature. If the budget is $2,000, the Earned Value is $1,300, and the CPI is 0.9, what is the Estimate to Complete?

   A. $800
   B. $630
   C. $700
   D. $1,300

2. As the manager of a software development project, you need to give senior management a report on the status of the project and a forecast for completion costs. If the project is budgeted at $100,000, the CPI is 0.8, and $20,000 of the budget has already

been spent, how much more money do you estimate you will spend to complete the project?

A. $105,000
B. $125,000
C. $145,000
D. $100,000

3. You are the Project Manager of a project budgeted at $240,000 that has shown no evidence of variance from Planned Value. Given a CPI of 1.2, what would you estimate to be the project's cost at completion?

A. $200,000
B. $120,000
C. $220,000
D. $250,000

4. Halfway through a project, it was determined that the cost and schedule variance were so high that the project needed a new baseline. An independent committee has determined that the cost of completing the remaining work is $8,000. If the team has already burned through $500 of the budget, what is the expected cost of the project at completion?

A. $8,000
B. $8,500
C. $500
D. $7,500

5. A construction project has the following parameters:

AC = $15,000; BAC = $60,000; EAC = $70,000; ETC = $55,000

Given this data, what amount of money is expected to be spent on the remainder of the project?

A. $40,000
B. $55,000
C. $70,000
D. $60,000

6. A Project Manager realizes that staff augmentation is a typical source of variance in projects in the organization. For a particular

project budgeted at $18,000, $10,000 has been earned back in value, while $8,000 has already been spent. The CPI is determined to be 1.25. What is the estimate of the cost at completion?

A. $14,400
B. $16,250
C. $12,000
D. $16,000

7. A Project Manager originally estimated a project's Budget at Completion to be $4,500, but the variances in the project were so high that the initial estimate was scrapped and the completion cost is now estimated at $8,500. What is the Variance at Completion?

A. (+) $4,500
B. (+) $8,500
C. (−) $4,000
D. (+) $13,000

8. You are the Project Manager for an IT project and have to determine the ETC. Past experience with IT projects for this organization leads you to believe that you will have similar variances in this project too. Given this information, what is the correct formula to use for Estimate to Complete?

A. $ETC = (BAC − EV) \div CPI$
B. $ETC = (BAC − EV)$
C. $ETC = (EAC − AC)$
D. $ETC = (EAC − AC) \div CPI$

9. A project had suffered a cost and a schedule setback. The Project Manager assumes that the variances are typical for this kind of project. Given a budget of $7,000, an Earned Value of $5,500, and a CPI of 0.6, what is the Estimate to Complete?

A. $1,500
B. $5,500
C. $7,500
D. $2,500

10. For a certain project, the Project Manager estimates the Budget at Completion to be $7,000. A recent Estimate to Complete was

$6,000, and $2,000 has already been spent. What is the Variance at Completion?

A. (+) $7,000
B. (−) $1,000
C. (+) $1,000
D. (−) $13,000

Exercise Answers

1. C

The key is that the variances were found to be atypical in nature and thus not likely to occur again. So the correct formula to use for ETC is

$$ETC = (BAC - EV)$$
$$ETC = 2,000 - 1,300 = \mathbf{\$700}$$

2. A

The estimate of how much more money must be spent to complete the project is the ETC. The correct formula to use is: ETC = EAC − AC (because EAC = ETC + AC and you are told that the actual cost is $20,000)

Step 1: Find the EAC.

$$EAC = BAC \div CPI = 100,000 \div 0.8 = 125,000$$

Step 2: Calculate the ETC.

$$ETC = EAC - AC = 125,000 - 20,000 = \mathbf{\$105,000}$$

3. A

To determine which formula to use, you have to look for clues in the question. The estimate of the cost at completion is the EAC. Another clue is that there have been no variances in the project thus far.

So, $$EAC = BAC \div CPI$$
$$EAC = 240,000 \div 1.2 = \mathbf{\$200,000}$$

4. B

Considering the fact that you are rebaselining and the committee has made a fresh estimate of the cost of the remaining work:

$$EAC = AC + ETC$$
$$EAC = 500 + 8,000 = \mathbf{\$8,500}$$

5. B

Of the various parameters given in the question, the amount of money expected to be spent on the remainder of the project is the Estimate to Complete, that is, **$55,000**.

6. A

Because variances caused by staff augmentation are considered typical, you can use the formula following to calculate the EAC:

$$EAC = AC + (BAC - EV) \div CPI$$

$$EAC = 8,000 + (18,000 - 10,000) \div 1.25$$

$$EAC = 8,000 + 6,400 = \$14,400$$

7. C

Variance at Completion is calculated by the formula VAC = BAC − EAC

$$VAC = 4,500 - 8,500$$

$$VAC = (-) \$4,000$$

8. A

Because the variances are expected to be typical of other IT projects in the organization, you should use the formula:

$$\mathbf{ETC = (BAC - EV) \div CPI}$$

9. D

The risk is considered to be typical, which means that it can occur again in the future. So this time you take the CPI into consideration. In this case,

$$ETC = (BAC - EV) \div CPI$$

$$ETC = (7,000 - 5,500) \div 0.6 = \$2,500$$

10. B

To find the EAC, use the formula EAC = AC + ETC.

$$EAC = 2,000 + 6,000 = \$8,000$$

Variance at Completion is calculated by the formula VAC = BAC − EAC

$$VAC = 7,000 - 8,000$$

$$VAC = (-) \$1,000$$

CHAPTER 4

Monitoring and Controlling

This chapter covers	1. Probability distribution
	2. Quality control tools

INTRODUCTION

The monitoring and controlling phase of a project is pivotal to ensuring that the project stays on track in terms of scope, schedule, performance, quality, and budget. The role of the Project Manager in this phase of the project is to use tools to constantly monitor progress, identify variances, and make adjustments.

A Project Manager skilled at proactively monitoring and measuring performance can boost the chances of success on the project. The actions taken during this phase are usually more aimed at ensuring that the project's deliverables conform to predefined quality standards. But actions should not be limited to just monitoring quality; they should also include comparing project performance with the schedule, cost, and scope baselines at regular intervals.

In the sections that follow, we will examine some of the tools and methods that can assist a Project Manager in monitoring and measuring different attributes of a project.

Another aspect of the monitoring and control phase is that it plays a vital role in ensuring the controlled and precise execution of the Project Management Plan created in the planning phase. Monitoring project execution allows the Project Manager to be proactive rather than reactive, and controlling helps address risks that threaten project success.

Probability Distribution

Probability distribution can help the Project Manager estimate budgets and schedules and perform resource allocation. It is also used as an input to tools that can simulate project schedules and generate numerous possible outcomes for a project based on the probability distribution function. Monte Carlo analysis is an example of one such simulation tool that is often used by Project Managers.

Quality Control Tools

Quality control charts extend the quality aspect of project management, allowing a Project Manager to systematically apply quality tools to analyze processes and conditions, review and collect data, and make decisions and inferences that affect quality.

4.1 PROBABILITY DISTRIBUTION

Main Concept

There is inherent variance in the value of any variable or project parameter that is sampled from a population. More often than not, the values of a certain variable will be different in every sample. The degree of variance depends on the sampling technique and the existence of any data integrity errors. But given this degree of variance, how do you determine the range of possible values for a certain variable (event)? The answer is the probability distribution.

Probability distribution, which is defined by a probability function, is used to illustrate the range within which the probability of a certain event will fall. It links each possible outcome of an event (variable or statistical experiment) with its probability. The range itself is limited by the statistically minimum and maximum values possible for that variable.

Probability distributions are broadly classified as discrete or continuous. In a discrete distribution, the variable in question can have only a finite number of possible values. A good example is a coin toss, with heads and tails as the only possible values. By contrast, in a continuous distribution, the variable can have a potentially unlimited number of values. A good example is a variable representing the age of every citizen in a country.

There are four types of continuous distributions, which are discussed in the following sections.

Beta Distribution

This kind of distribution curve (See Fig. 4.1) can take a wide range of shapes and represents the uncertainty or randomness of the probability of an event. The data points on the curve are not equally distributed on either side of the mean. Beta distributions model events that exhibit random variation, but are restricted within a range, i.e., between a maximum and minimum value. An example of a beta distribution is the grades received by a graduate class on their mid-term paper. All grades are between 0 and 100, but they may be distributed randomly on either side of the mean.

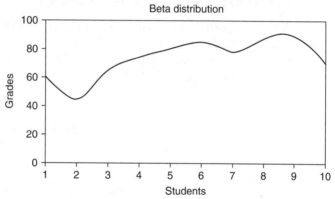

Figure 4.1 Beta distribution.

Triangular Distribution

Triangular distribution is used to model events where the most likely outcomes/ values are known. As shown in Fig. 4.2, the most likely values are clustered,

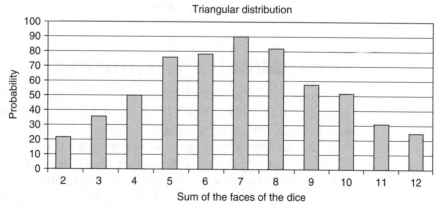

Figure 4.2 Triangular distribution.

indicating a higher probability of occurrence. The probability declines on either side of the most likely outcomes. An example is rolling a pair of dice and tracking the sum of the values on the faces of the dice. The probability that the sum of the faces of the dice will be 2 or 12 is far lower than the probability that it will be 5, 6, or 7. You will roll a sum of 2 or 12 only when you roll a combination of (1,2), (2,1), or (6,6). But you are much more likely to roll a combined 7 because there is a greater number of favorable combinations: (1,6), (6,1), (2,5), (5,2), (3,4), and (4,3).

Uniform Distribution

A uniform distribution indicates that each outcome has the exact same probability of occurrence (See Fig. 4.3). An example is flipping a coin: no matter how many times you flip it, the probability of getting a head or a tail is the same for every flip.

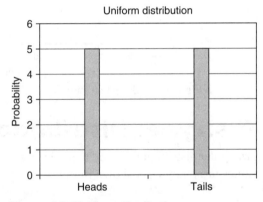

Figure 4.3 Uniform distributions.

Normal/Bell Distribution

This kind of distribution curve, which is also referred to as the bell curve, is symmetric on either side of the center point, which is the mean (average). An example of a bell curve is a graph plotting the age of all persons attending college. The curve is symmetric in nature with the average between 20 and 40. The probability that the college students are 20 to 40 years old is greater than the probability that they are 10 to 20 or 40 to 50 years old. As the curve above indicates, there is almost zero chance that they are aged less than 10 or greater than 50 years (See Fig. 4.4).

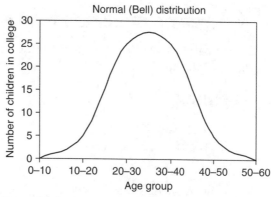

Figure 4.4 Normal distribution.

The term "Sigma" denotes a measure of the standard deviation or variation of a statistical population on a normal curve. "Six Sigma" is a quality improvement practice pioneered by Motorola primarily to reduce and eliminate defects in its manufacturing processes, assuming the process conforms to a normal distribution. Six Sigma measures six standard deviations between the mean and the specified control limit of the process being measured, with the expectation that 99.9985% of the population would fall within this range.

Probability Distribution Problems on the PMP Exam

TYPES OF PROBLEMS TO EXPECT

There are four types of probability distribution problems that you can expect to encounter on the PMP exam:

(1) Determine the type of distribution used.
(2) Determine the range of estimates given the type of distribution.
(3) Determine the mean/average given the type of distribution.
(4) Determine the variance given the type of distribution.

DIFFICULTY LEVEL

These problems are relatively easy. Once you work through the exercise problems, you will definitely have a good grasp of the subject.

NUMBER OF PROBLEMS TO EXPECT

There will be a maximum of one question on probability distribution on the PMP exam.

FORMULAS

For beta distributions:

Mean $= (P + 4 \times M + O) \div 6$

Variance $= ((P - O) \div 6)^2$

Standard deviation $= \sqrt{\text{variance}} = ((P - O) \div 6)$

For triangular distributions:

Mean $= (P + O + M) \div 3$

Variance $= (O^2 + P^2 + M^2 - O \times P - P \times M - O \times M) \div 18$

Standard deviation $= \sqrt{\text{variance}}$

INSIDER TIPS

• Identifying the type of distribution is key to knowing which set of formulas to apply. Some questions might refer to a distribution by an alternate name, such as *"bell-curve"* or *"Shewhart"* for a normal distribution.

SAMPLE SOLVED PROBLEMS

1. A Project Manager is using the three-point estimating technique to estimate the duration of a scheduled activity. Given the information below, which type of distribution was assumed for the project?

   Pessimistic $= 100$ days

   Optimistic $= 75$ days

   Most likely $= 85$ days

   Mean $= 86.67$ days

   Step 1: First check to see if this is a beta distribution. Use the formula to find the mean.

   Mean $= (P + 4 \times M + O) \div 6$

   Mean $= (100 + 4 (85) + 75) \div 6 = 85.83$

   Because the mean does not equal 86.67, which is the given mean, this is not a beta distribution.

Step 2: Next, check to see if this is a triangular distribution. Use the formula to find the mean.

Mean = (P + O + M) ÷ 3

Mean = 100 + 75 + 85 ÷ 3 = 86.67 days

Because the mean matches the data given, this must be a triangular distribution.

2. A Project Manager is using the three-point estimating technique to estimate the duration of a scheduled activity. Given the information below, which type of distribution was assumed for the project?

Pessimistic = 100 days

Optimistic = 76 days

Most likely = 85 days

Variance = 24.5 days

Step 1: First check to see if this is a beta distribution. Use the formula to find the variance.

Variance = $((P - O) \div 6)^2$

Variance = $((100 - 76) \div 6)^2 = 16$

Because the given variance does not equal 16, this cannot be a beta distribution.

Step 2: Next, check to see if this is a triangular distribution. Use the formula to find the variance.

Variance = $(O^2 + P^2 + M^2 - O \times P - P \times M - O \times M) \div 18$

Variance = $(76^2 + 100^2 + 85^2 - 76 \times 100 - 100 \times 85 - 76 \times 85) \div 18$

Variance = $(5,776 + 10,000 + 7225 - 7,600 - 8,500 - 6,460) \div 18$

Variance = $441 \div 18 = 24.5$

Because the variance matches the data given, this must be a triangular distribution.

EXERCISE PROBLEMS

1. Which probability distribution is used to measure Six Sigma?

    A. Beta distribution
    B. Normal distribution
    C. Triangular distribution
    D. Uniform distribution

2. Which type of distribution model indicates that each outcome has the same probability of occurrence as every other outcome?

   A. Normal distribution
   B. Beta distribution
   C. Triangular distribution
   D. Uniform distribution

3. A Project Manager is working with his team lead to estimate the duration of a task to develop a software module to allow users to transfer money between bank accounts. The team lead gives an optimistic estimate of 500 man-hours, a pessimistic estimate of 560 man-hours, and a most likely estimate of 530 man-hours. Taking beta distribution into account, what will be the range for the Project Manager's estimate?

   A. Between 500 and 600 man-hours
   B. Between 600 and 650 man-hours
   C. Between 520 and 540 man-hours
   D. Between 500 and 530 man-hours

4. A Project Manager is trying to determine the testing schedule for a project. The QA lead gives her a pessimistic estimate of 360 hours, an optimistic estimate of 300 hours, and a most likely estimate of 330 hours. If the variance is 100, which type of distribution did the Project Manager assume while calculating the estimate for the project?

   A. Normal distribution
   B. Beta distribution
   C. Triangular distribution
   D. Uniform distribution

5. A Project Manager is trying to determine the QA schedule for a certain activity. The QA lead gives her a pessimistic estimate of 500 hours, an optimistic estimate of 420 hours, and a most likely estimate of 460 hours. If the variance is 266.67, which type of distribution did the Project Manager assume for this activity?

   A. Normal distribution
   B. Beta distribution
   C. Triangular distribution
   D. Uniform distribution

6. A Project Manager is trying to determine the amount of time it will take to regression test a module of the product. The QA lead gives an optimistic estimate of 120 man-hours, a pessimistic estimate of 150 man-hours, and a most likely estimate of 130 man-hours. Taking beta distribution into account, what schedule range should the Project Manager estimate?

    A. Between 127 and 137 man-hours
    B. Between 100 and 150 man-hours
    C. Between 120 and 157 man-hours
    D. Between 107 and 150 man-hours

7. A Project Manager is trying to determine the number of defects in a product. The QA Manager gives an optimistic estimate of 50 defects, a pessimistic estimate of 80 defects, and a most likely estimate of 65 defects. Taking beta distribution into account, what is the average number of defects the module is shipped with?

    A. 80 defects
    B. 65 defects
    C. 50 defects
    D. 100 defects

8. A Project Manager is trying to determine the number of inherent defects in a product. The QA Manager gives an optimistic estimate of 50 defects, a pessimistic estimate of 80 defects, and a most likely estimate of 65 defects. Taking triangular distribution into account, what is the average number of defects the module is shipped with?

    A. 65 defects
    B. 50 defects
    C. 130 defects
    D. 100 defects

9. A Project Manager is trying to determine the amount of time it will take to analyze a Web site that allows prospective employees to submit job applications. The Business Analyst gives an optimistic estimate of 90 man-hours, a pessimistic estimate of 120 man-hours, and a most likely estimate of 100 man-hours. Taking beta distribution into account, what will be the range of the Project Manager's estimate?

A. 102 ± 25 days

B. 102 ± 5 days

C. 200 ± 25 days

D. 200 ± 5 days

10. A consultant has given an optimistic estimate of 30 man-hours, a pessimistic estimate of 39 man-hours, and a most likely estimate of 33 man-hours. As the Project Manager, you decide to take triangular distribution into account. What is the variance for the consultant's estimate?

A. ± 3 man-hour

B. ± 3.5 man-hours

C. ± 4 man-hours

D. ± 4.5 man-hours

Exercise Answers

1. **B**

One of the features of a normal curve is that 99.7% of the data points lie within 3 standard deviations of the mean (i.e., on either side). This equates to Six Sigma.

2. **D**

Uniform distribution indicates that each outcome has the same probability of occurrence as every other outcome.

3. **C**

For a beta distribution, the mean is calculated using the formula:

Mean $= (P + 4 \times M + O) \div 6$

So in this case,

Mean $= (560 + 530 \times 4 + 500) \div 6 = 530$

For a beta distribution, standard deviation is calculated using the formula:

Std Deviation $= ((P - O) \div 6) = 10$

So the PM needs to prepare for 530 ± 10 hours, or for **520 to 540 hours**

4. **B**

Assuming beta distribution,

Variance $= ((P - O) \div 6)^2 = (360 - 300) \div 6)^2 = 100$

Because this matches the given data, the PM must have assumed a **beta distribution** for the project.

5. **C**

Assuming beta distribution,

Variance $= ((P - O) \div 6)^2 = (500 - 420) \div 6)^2 = 177.78$

Because this does not match the given variance, it is not a beta distribution.

Assuming triangular distribution,

Variance $= (O^2 + P^2 + M^2 - O \times P - P \times M - O \times M) \div 18$

Variance $= 266.67$, which does match the given data.

6. **A**

For a beta distribution:

Mean $= (P + 4 \times M + O) \div 6$

Mean $= (150 + 130 \times 4 + 120) \div 6 = 131.67$ (rounded to 132 days)

For a beta distribution, standard deviation is calculated using the formula:

Std Deviation $= ((P - O) \div 6) = 5$

So the PM needs to prepare for 132 ± 5 days or for **127 to 137 hours.**

7. **B**

For a beta distribution, the mean is calculated using the formula:

Mean $= (P + 4 \times M + O) \div 6$

Mean $= (80 + 65 \times 4 + 50) \div 6 = $ **65 defects**

8. **A**

For a triangular distribution, the mean is calculated using the formula:

Mean $= (P + O + M) \div 3$

Mean $= (80 + 65 + 50) \div 3 = $ **65 defects**

9. **B**

For a beta distribution, the mean is calculated using the formula:

Mean $= (P + 4 \times M + O) \div 6$

Mean $= (120 + 100 \times 4 + 90) \div 6 = 101.67$ (rounded to 102 days)

For a beta distribution, standard deviation is calculated using the formula:

Std Deviation $= ((P - O) \div 6) = 5$

So the PM needs to prepare for **102 ± 5 days**

10. **B**

Variance $= (O^2 + P^2 + M^2 - O \times P - P \times M - O \times M) \div 18$

Variance $= (30^2 + 39^2 + 33^2 - 30 \times 39 - 39 \times 33 - 30 \times 33) \div 18$

Variance $= (900 + 1521 + 1089 - 1170 - 1287 - 990) \div 18$

Variance $= 63 \div 18 = 3.5$

So the variance is **± 3.5 man-hours**

4.2 QUALITY CONTROL TOOLS

Main Concept

Quality control tools provide the Project Manager with the ability to analyze a problem and get to the root cause and fix it, instead of just implementing a quick fix that may not be a long-term solution. Many different quality control tools are available to a Project Manager depending on the situation, the type of quality problem to evaluate, and the decisions to be made. Examples of quality control tools include control charts, Pareto charts, histograms, bar charts, cause-and-effect diagrams, run charts, flow charts, check sheets, inspections, scatter diagrams, and statistical sampling.

Two quality control tools—Pareto charts and control charts—involve a bit of math computation that may be tested on the PMP exam, so those are the only ones discussed here in detail. The PMP exam will also include other general conceptual questions about quality control charts, so we have provided a brief overview of those in the following section.

Control Charts

A control chart is a statistical tool used to visually depict the behavior of a process over time. It is also known as a Shewhart chart (named after its inventor, the American engineer and statistician Walter A. Shewhart). Project

Managers can use this tool to examine the process for undesired variations and fluctuations, indicating loss of control. Control charts describe a predetermined range within which the process is expected to remain and the average value. Any outliers (data points beyond the defined range/limits) warrant additional investigation. Apart from outliers, the Project Manager should also look for specific patterns in the variation of the process, even if it is within the limits defined. Observations that are beyond limits or unlikely variation patterns within limits are referred to as a *special cause* and need to be examined closely.

Another way to interpret a control chart is to use the *rule of seven*. This rule states that a process is out of control if there are either at least seven data points in a row above or below the average, or seven consecutive data points with an upward or downward trend.

The chart shown in Fig. 4.5 is a typical control chart illustrating the time required to load a particular Web site. The response time is different on different occasions (data points). This chart does not indicate the limits. However, a control chart does not visually denote the standard deviation of the process (i.e., it does not represent the level of Sigma).

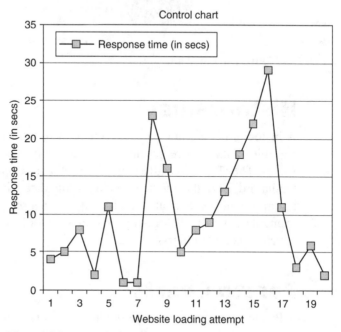

Figure 4.5 Control chart.

Bar Charts

Bar charts (or bar graphs) are used most commonly for comparative analysis. A bar chart is a style of graph that uses rectangular bars to indicate the values represented.

The bar chart shown in Fig. 4.6 illustrates the quarterly sales figures for a company.

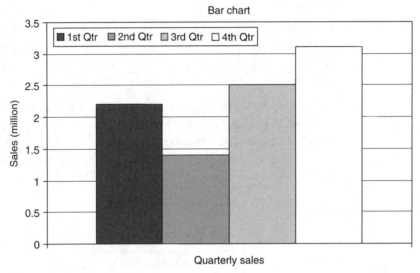

Figure 4.6 Bar chart.

Histograms

A histogram is a graphical representation of the frequency distribution of a variable. That is, it represents how frequently the variable assumes certain values. The chart is usually in the form of a bar chart. The bars are in no particular order, but they do help in identifying data with the maximum and minimum frequency and also in identifying a distribution pattern: uniform, triangular, or bell. Histograms also serve as tools to help in identifying symmetric data and outliers.

Pareto Charts

A Pareto chart (named for the Italian economist Vilfredo Pareto) is like a histogram (bar chart), except that the data points are plotted in descending order

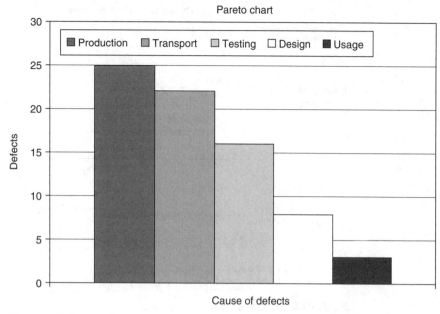

Figure 4.7 Pareto chart.

of value. This value could be the frequency of occurrence of a certain event or just a variable value. By arranging the data in the chart in the descending order, the Pareto chart tries to highlight the more frequent or more important set of values.

Pareto charts also extend Pareto's "80-20 principle," which states that 80% of the problems (most frequent) are caused by 20% of the underlying issues.

The Pareto chart shown in Fig. 4.7 focuses on the most frequent cause of product defects.

Cause and Effect Diagrams

Also known as an *Ishikawa* or *fishbone* diagram, this tool is primarily used in root cause analysis. It is a critical visual aid to help identify the root cause of quality issues and relationships between different variables.

The fishbone diagram shown in Fig. 4.8 analyzes the primary and secondary causes of low customer satisfaction following the completion and delivery of a product/project.

Cause and effect diagram

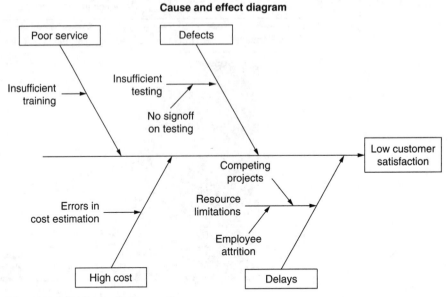

Figure 4.8 Ishikawa/fishbone diagram.

Run Charts

A run chart is time sensitive and is primarily used to capture data and measure performance over a period of time. The chart also helps identify trends and patterns in the data and hence is also referred to as a trend chart. A run chart is a control chart without the specification limits.

The run chart shown in Fig. 4.9 shows the progression of the cost incurred on a project over an entire year, month to month.

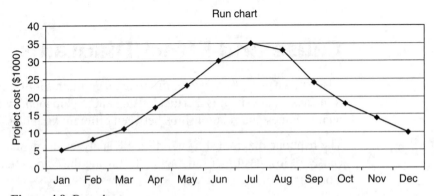

Figure 4.9 Run chart.

Scatter Diagrams

A scatter diagram is a tool designed to analyze the relationship (if any) between two variables. The two variables are plotted on either axis of the graph and the variable values are plotted as data points. The resultant data points could indicate trend lines and some correlation between the two variables. A caveat to using scatter diagrams is that even though the diagram may indicate that the two variables are causally related, that may not be the case in reality. What the diagram does do is give a clue to possible relations between two variables.

The scatter diagram shown in Fig. 4.10 plots the salary of employees of a company against the calculated attrition rate. The downward trend line indicates that as the salary increases, the attrition rate decreases. The chart suggests a relationship between these two variables, but it doesn't prove that a relationship actually exists.

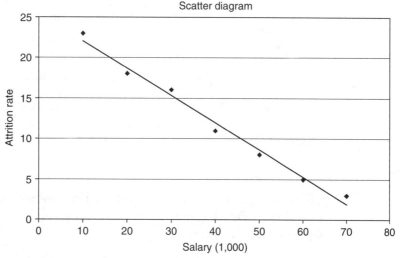

Figure 4.10 Scatter diagram.

Statistical Sampling

Statistical sampling is the process of collecting samples and analyzing them in order to make inferences about an entire population. The sample size and sampling technique are key factors that determine the accuracy and effectiveness of statistical sampling.

Flow Charts

A flow chart is a tool that can be used not only for quality control, but also for other stages like planning and designing. A flow chart documents the flow of a process or a sequence of steps involved in a process. It promotes better understanding of the process and helps bring out any underlying issues or problems with the process. With respect to quality control, a review of a flow chart can reveal defects in a process and alternate ways to improve quality.

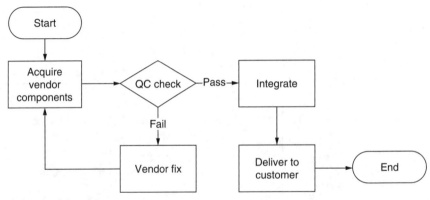

Figure 4.11 Flow chart.

The flow chart shown in Fig. 4.11 documents the process of getting components from vendors, testing them individually, and then integrating them into a product for the customer.

Inspection

Inspection is an important quality tool that is used in the monitoring and controlling phase. It is a key component of quality assurance testing. The inspection of the results of a task always follows the execution of the task. While inspection does not prevent a defect, it prevents the defect from being shipped to the customer. Inspection ensures defect identification before customer delivery.

Check Sheets

A check sheet (also referred to as a checklist) is a structured data collection tool that is aimed at capturing data at its source (See Fig. 4.12). It is typically used for repetitive tasks or for collecting data about the frequency of occurrence of a certain event, such as a defect.

	Check sheet—first quarter metrics—2008			
	Jan	Feb	Mar	Total
Suggestions	I	IIII	II	7
Positive feedback	II	卌	卌 II	14
Complaints	卌 II	II	IIII	13
Errors	卌	卌	II	12
TOTAL	15	16	15	46

Figure 4.12 Check sheet.

Quality Control Tools Problems on the PMP Exam

TYPES OF PROBLEMS TO EXPECT

Problems on the PMP exam in the area of quality control tools that involve math are almost always related to Pareto or control charts. Typical PMP questions take the following forms:

(1) Find upper/lower limits and outliers in a control chart.
(2) Determine if a process is out of control using a control chart.
(3) Use the rule of seven to identify special causes.
(4) Interpret a Pareto chart.

DIFFICULTY LEVEL

These problems do not require much practice. Be aware of the purpose of each tool and work through the exercise questions.

NUMBER OF QUESTIONS TO EXPECT

On the PMP exam, expect a maximum of three questions on quality control tools. The questions could be math-based, but they could also test your understanding of the concepts discussed above.

SAMPLE SOLVED PROBLEMS

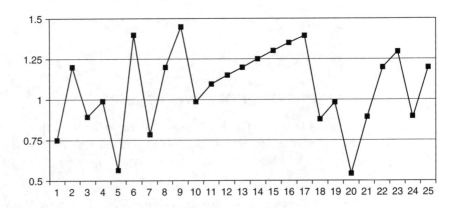

1. The preceding chart is a control chart with specification limits of 1.25 and 0.75. Given this data, answer the following questions:

   a. **Which data points of the process are out of control?**

      To answer this question, you need to review the control chart and find data points that are above the upper limit and below the lower limit. In this case it is data points 6, 9, 15, 16, 17, 23 that are above the upper limit and data points 5, 20 that are below the lower limit. These outliers are the data points that indicate that the process is out of control.

b. **Which data points indicate the presence of a "special cause" ("rule of seven")?**

A quick visual scan indicates that it is data points 10 to 17. Note that these 8 consecutive points exhibit a continuous upward trend. Recall that a 'rule of seven' exists if one of the four conditions exists:

1. At least seven consecutive points below the mean
2. At least seven consecutive points above the mean
3. At least seven consecutive points with an upward trend
4. At least seven consecutive points with a downward trend

EXERCISE PROBLEMS

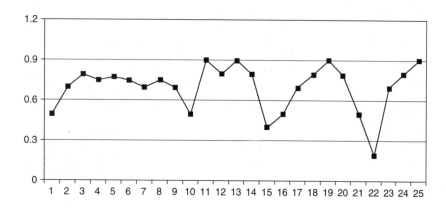

1. Based on the preceding chart, answer the following questions:

a. Based on the upper and lower range given above, what is the company's quality standard?

   A. Six Sigma
   B. Three Sigma
   C. You cannot tell the level of Sigma from the control chart.
   D. Two Sigma

b. At what point is the process out of control?

   A. Data point 10
   B. Data point 22
   C. Data point 15
   D. Data point 25

c. Which data points indicate a rule of seven?

A. Data points 10 to 16
B. Data points 2 to 8
C. Data points 15 to 21
D. Data points 1 to 7

d. Which occurrences of the process indicate an assignable cause?

A. Data points 2, 3, 4, 5, 6, 7, 8, 9 and 22
B. Data points 2, 3, 4, 5, 6, 7, 8 and 9
C. Data points 2, 3, 4, 5, 6, 7 and 8
D. All data points

e. What is the upper control limit?

A. 0.9
B. 0.6
C. 0.3
D. 0

f. What is the lower control limit?

A. 0.9
B. 0.6
C. 0.3
D. 0

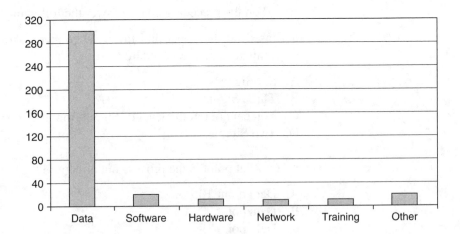

2. The preceding Pareto chart indicates all incoming customer service calls classified by the problem the call relates to. Based on Pareto analysis, which area should the Project Manager focus on?

   A. Data
   B. Hardware
   C. Software
   D. All except Data

3. What principle does the Pareto chart follow?

   A. 80-20 principle (80% of problems are due to 20% of the causes.)
   B. 70-30 principle (70% of problems are due to 30% of the causes.)
   C. 50-50 principle (50% of problems are due to 50% of the causes.)
   D. 20-80 principle (20% of problems are due to 80% of the causes.)

4. If you wish to observe the number of defects over time, which quality control tool would you use?

   A. Pareto chart
   B. Histogram
   C. Run chart
   D. Scatter diagram

5. If you have a huge application and do not have the time and manpower to test all the different functions, which quality control technique would you use?

   A. Statistical sampling
   B. Flow charting
   C. Inspection
   D. Cause-and-effect diagram

Exercise Answers

1.
   a. **C**
      You cannot tell the level of Sigma from a control chart.

   b. **B**
      Data point 22.

c. **B**

The occurrences between **data points 2 to 8** indicate the rule of seven. A "rule of seven" exists if one of these four conditions exists:

1. At least seven consecutive points below the mean
2. At least seven consecutive points above the mean
3. At least seven consecutive points with an upward trend
4. At least seven consecutive points with a downward trend

d. **B**

These data points are consecutive and lie on the same side of the mean. This consecutive clustering suggests an assignable cause and not a random event. An assignable cause is an identifiable or a specific cause of variation in a process.

e. **A**

An upper limit indicates the acceptable limit above the mean beyond which any data points will be considered as outliers.

f. **C**

A lower limit indicates the acceptable limit below the mean beyond which any data points will be considered as outliers.

2. **A**

The area to focus on is **data** because that is the area that is generating the most customer service calls.

3. **A**

The Pareto principle is the **80-20 principle**, which suggests that 80% of problems in a project are usually generated by 20% of the causes.

4. **C**

The **run chart** is plotted against time and is a perfect candidate for tracking defects over time and analyzing any trends or patters in it.

5. **A**

In the absence of resources and time to perform thorough testing, it is prudent to carefully sample the available data and test just the samples. This is also referred to as **statistical sampling**. There is some margin for error when you use sampling for testing, and the margin is dependant on the sampling technique.

CHAPTER 5

Closing

This chapter covers	1. Statistical Concepts • Mean • Median • Mode • Standard deviation • Variance

INTRODUCTION

The closing phase of the Project Management Cycle helps tie up all loose ends and bring about a complete and official closure to the project. This is the point at which any contracts with vendors are closed, payments are made, resources are released, and project documents such as the lessons learned are archived. But too often a Project Manager fails to devote enough time and resources to the closing phase, and hence it is actually the most overlooked phase of a project. Part of this neglect comes from the fact that by this time, the product of the project has already been completed, so teams usually tend to overlook the value of activities such as archiving lessons learned for a project or documenting performance metrics.

Statistical concepts can be applied at any stage of the project to determine a variety of variables and metrics. More often than not they are used to evaluate a performance metric for a project or to exercise quality control. We have included this introduction to statistical concepts in this chapter, focusing on closing in a project.

5.1 STATISTICAL CONCEPTS

Statistical concepts such as the mean, median, mode, and standard deviation are basic terms that a Project Manager should thoroughly understand. A good grasp of these concepts can help the Project Manager apply them to real situations in the project for performance measurements. Statistics help a Project Manager measure the central tendency of a given data set. Such measurements can then be used to assess the quality of the results of the project and to analyze problems and defects.

Main Concept

This section will discuss in detail statistical terms such as mean, median, mode, standard deviation, variance, and range. It will also present two data analysis techniques that make use of these statistical terms and are also pertinent to the PMP exam.

The data analysis techniques are the following:

(1) **Deriving the central tendency of a data set.** This involves finding a number or value to earmark or characterize the given data set.
(2) **Deriving the dispersion of a data set.** This involves analyzing variability from and proximity to the central tendency of the data set.

Measures of Central Tendency: Mean, Median, and Mode

A common way to characterize a data set is to calculate a measure of its so-called "central tendency." This means finding a single number that can represent or describe all the numbers in the data set. The three most commonly used measures of central tendency are the mean, median, and mode. Calculating these measures for the data generated by your project can give you a quick "snapshot" of the project's performance. You can also use the mean, median, and mode to measure how your project stacks up against other projects with similar scope.

MEAN/AVERAGE

On the PMP exam, the terms *mean* and *average* always refer to the arithmetic mean. The arithmetic mean is one measure of the central tendenccy of a given data set. It is the sum of all the data points in the set divided by the number of data points (also called *observations*).

A simple example is a data set of test scores from 10 PMP test takers: {72, 85, 33, 85, 97, 56, 67, 85, 90, and 94}. To compute the mean, first find the sum of all 10 scores, which is 764. Then divide that sum by the number of scores: $764 \div 10 = 76.4$. This is the arithmetic mean.

MEDIAN

To compute the median, first arrange the data points (observations) in sequence in ascending order of magnitude. Then find the value of the data point that occurs in the middle of the sequence. When you have an odd number of data points, the middle value is clearly evident. When you have an even number of data points, the median is the arithmetic mean of the two numbers in the middle.

For example, here is a data set with an odd number of values: {72, 85, 33, 85, 97, 56, 67, 85, 90}. Start by arranging the values in ascending order of magnitude: {33, 56, 67, 72, 85, 85, 85, 90, 97}. The value of the data point in the middle is 85, which is the median.

Now consider a data set with even number of values: {72, 85, 33, 88, 97, 56, 67, 87, 90, 98}. Arrange the values in ascending order of magnitude: {33, 56, 67, 72, 85, 87, 88, 90, 97, 98}. The two middle data points in this sequence are the fifth value (85) and the sixth value (87). The arithmetic mean of these two values is 86, which is the median.

MODE

In statistics, the term mode denotes the value that occurs most frequently in a given data set. For example, consider the following data set: {72, 85, 33, 85, 97, 56, 67, 85, 90, 98}. The value that occurs most frequently is 85, so that is the mode. (In cases where each value in the data set occurs only once, or where two or more values occur with the same frequency, there is no mode for that data set.)

Dispersion of the Data: Standard Deviation, Variance, and Range

Another way to characterize a data set is to measure its dispersion, or deviation, from the central tendency. Measuring dispersion allows you to gauge the variability of the data and also to identify any outliers in it. The statistical tools used to measure dispersion are the *standard deviation* and the *variance*. Another tool, called the *range*, measures the difference between the maximum and the minimum values in the data set.

STANDARD DEVIATION

In statistical terms, standard deviation shows deviation from the mean. The closer a value is to the mean, the more "normal" the value. A low standard deviation indicates that the data is close to the mean. A high standard deviation can imply that there are data outliers that need to be analyzed.

The standard deviation of a data point is given by the difference between the value of the data point and the mean of the entire data set. i.e., standard deviation = (data point value – mean). Ignore any negative sign you get for the standard deviation, since what you are interested in is just the magnitude of the difference from the mean, not whether it is positive or negative. Please note that the formula above works only for calculating the standard deviation of a single data point amongst a data set.

Given an entire set of data points, the standard deviation of this entire set is calculated by performing the following steps:

- First, compute the square of the difference between each data point and the mean.
- Second, determine the average of the values computed in the step above. This is also the variance of the data set.
- Third, take the square root of the average determined in the step above. This will represent the standard deviation of the entire data set.

If the data points represent the estimated duration for completing activities on a project and we are to use the PERT method to evaluate the expected duration of an activity or the project as a whole, the calculation of the standard deviation uses a slightly different approach as explained below (although it is fundamentally along the same lines as discussed above).

The PERT method assumes a beta probability distribution for activity time estimates and estimates are provided in three forms—optimistic (O), pessimistic (P) and most likely (M). The standard deviation in this case for the estimated time for an activity is given by: $(P - O) \div 6$.

When you are given an entire data set, i.e., an entire set of estimates for each activity in the project, the standard deviation of the expected duration of the project can be determined as follows:

- First, compute the variance of each activity, given as $[(P - O)/6]^2$.
- Second, add all the activity variances to get the project variance.
- Third, take the square root of the project variance determined in the step above. This will represent the standard deviation for the project and its expected duration.

In a normal distribution (often called a "bell curve" from its shape when the data is graphed):

- 68% of the data points in the set are ± one standard deviation from the mean.
- 95% of the data points in the set are ± two standard deviations from the mean.
- 99% of the data points in the set are ± three standard deviations from the mean.

VARIANCE

Variance is another measure of the dispersion of data. You compute the variance by squaring the standard deviation. Knowing how to find the variance will help you on the PMP exam if you are asked to find the standard deviation of an entire data set or group. Whenever you want to find the standard deviation of a group of data points, you cannot just add the standard deviations of the individual data points. You will need to calculate the variance of each data point then add the variances together. Finally, the square root of the total variance gives you the total standard deviation for the data set.

The variance of a data point is given by the square of the difference between the value of the data point and the mean of the entire data set. i.e., Variance = (data point value – mean)^2. Please note that the formula above works only for calculating the variance of a single data point amongst a data set.

Given an entire set of data points, the variance of this entire set is calculated by performing the following steps:

- First, compute the square of the difference between each data point and the mean.
- Second, determine the average of the values computed in the step above. This is also the variance of the data set.

The PERT method assumes a beta probability distribution for activity time estimates and estimates are provided in three forms—optimistic (O), pessimistic (P) and most likely (M). When you are given a set of activities for a project, i.e., an entire set of estimates for each activity in a project, the total project variance is determined as follows:

- First, compute the variance of each activity, given as $[(P - O)/6]^2$.
- Second, add all the activity variances to get the project variance.

RANGE

The range of a given data set is the difference between the minimum value and the maximum value in the set. For example, when you make a rough order magnitude cost estimation, the error margin is −50% to +100%. So that gives a range of $100 - (-50) = 150$.

Statistical Concepts Problems on the PMP Exam

TYPES OF PROBLEMS TO EXPECT

On the PMP exam, you can expect several types of problems based on the concepts in this chapter:

(1) Computing the arithmetic mean
(2) Computing the median
(3) Computing the mode
(4) Computing the standard deviation using pessimistic and optimistic estimates
(5) Computing the standard deviation using variance
(6) Computing the variance

DIFFICULTY LEVEL

These problems are fairly easy. Expect very simple problems related to statistics.

NUMBER OF PROBLEMS TO EXPECT

On the PMP exam, expect a maximum of one problem based on the concepts covered in this chapter

FORMULAS

- **Mean** Add all the data points in the data set, and then divide the sum by the number of data points.
- **Median** Arrange all of the data points in sequence based on ascending order of magnitude. If there is an even number of data points, the median is the arithmetic average of the middle two data points. If there are an odd number of data points, the median is the middle data point.
- **Mode** The mode is the data value that occurs most frequently in a given data set.
- **Standard Deviation** Standard deviation (SD) of a single data point in a given data set is

$$\text{SD (data point)} = (\text{data point value} - \text{mean of data set})$$

When we are using the PERT method, the formula to use is

$$SD = (P - O) \div 6$$

where P is a pessimistic estimate and O is an optimistic estimate.

- **Variance**   Variance (V) of a single data point in a given data set is

$$V \text{ (data point)} = (\text{data point value} - \text{mean of data set})^2$$

When we are using the PERT method, the formula to use is

$$\text{Variance (V)} = [(P - O) \div 6]^2$$

where P is a pessimistic estimate and O is an optimistic estimate.

- **Range**   The range of a data set is the magnitude of the difference between the maximum and minimum value amongst all data points in the set.

INSIDER TIPS

- The key is to make the distinction between calculating these statistical parameters for a data point or a set using the traditional approach or having to use the formulas for the PERT method.
- Remember that when calculating the standard deviation for an entire data set or group, do not just add the individual standard deviations.
- Please note that the most likely estimate is not used in computing the standard deviation and variance using the PERT method.

SAMPLE SOLVED PROBLEMS

1. A project has five activities defined in it. The cost of these activities is estimated at $4,000, $2,000, $6,000, $5,000, and $8,000. What is the estimated average cost of the project and what is the estimated standard deviation for the total cost of the project?

To compute the mean, count the number of data points. In this case, it is 5.

Mean = (4,000 + 2,000 + 6,000 + 5,000 + 8,000) ÷ 5 = $5,000

To compute the standard deviation for the total project cost, you need to go through the steps given below:

Step 1: Compute the average of the squares of the difference between the mean and each data point value.

$$\text{Sum of squares} = (4{,}000 - 5{,}000)^2 + (2{,}000 - 5{,}000)^2 + (6{,}000$$
$$- 5{,}000)^2 + (5{,}000 - 5{,}000)^2 + (8{,}000 - 5{,}000)^2$$

$$\text{Average of squares} = \text{sum of squares} \div 5 = 4{,}000{,}000$$

Step 2: Compute the standard deviation

$$\text{Standard deviation} = \text{square root (average of squares)} = \$2{,}000$$

EXERCISE PROBLEMS

1. You are the Project Manager for a software development project that involves working with three different vendors on contracts to deliver specific modules. The cost of these contracts is estimated to be $200,000, $350,000, and $350,000, respectively. Given this information, what is the average cost of procurement for this project?

   A. $300,000
   B. $350,000
   C. $200,000
   D. $190,000

2. The test scores of a sample set of PMP test takers are 99, 98, 56, 89, 96, 79, 80, 87, 91, 88, 95. What is the median test score?

   A. 99
   B. 95
   C. 89
   D. 56

3. The optimistic schedule estimate for a certain project is 22 days, the pessimistic schedule estimate is 40 days, and the most likely schedule estimate is 35 days. What is the standard deviation of the project schedule?

   A. 3 days
   B. 5 days
   C. 4 days
   D. 2 days

4. As a Project Manager, you are adopting the PERT method for determining the expected duration of an activity. The activity has an optimistic estimate of 20 days, a pessimistic estimate of 32 days, and the most likely estimate is 25 days. What is the variance of these schedule estimates for this activity?

   A. 9 days
   B. 25 days
   C. 4 days
   D. 6 days

5. The following data set represents a set of PMP test scores: {95, 75, 70, 95, 50, 75, 90, 80, and 75}. What is the mode of this data set?

   A. 95
   B. 50
   C. 70
   D. 75

Project Task	Standard Deviation in Weeks
Task A	2 weeks
Task B	3 weeks
Task C	6 weeks

6. What is the total standard deviation for the project's duration, given that the individual tasks in the project have standard deviations as given below:

   A. 3 weeks
   B. 6 weeks
   C. 7 weeks
   D. 5.5 weeks

Exercise Answers

1. **A**

   The procurement costs for the three contracts with vendors are listed as: $200,000, $350,000, and $350,000.

   $$\text{Total procurement cost} = \$900,000$$
   $$\text{Average cost of a procurement} = 900,000 \div 3 = \textbf{\$ 300,000}$$

2. **C**

There are 11 data points. Arranging them in ascending order you get $\{56, 79, 80, 87, 88, 89, 91, 95, 96, 98, 99\}$. So the median (the middle data point) is **89**.

3. **A**

The usage of optimistic, pessimistic, and most likely estimates tells you that you need to use the PERT technique to evaluate the standard deviation

$$\text{Standard deviation} = (P - O) \div 6.$$

$$\text{Standard deviation} = (40 - 22) \div 6 = \textbf{3 days}$$

4. **C**

Using the PERT method, Variance is calculated using the formula:

$$\text{Variance} = [(P - O) \div 6]^2$$

$$\text{Variance} = [(32 - 20) \div 6]^2 = \textbf{4 days}$$

Note that the most likely estimate is not used in the determination of the variance. It is just additional information provided that isn't relevant to calculating the variance.

5. **D**

The mode of a given data set is the value that occurs most frequently in the set. For this data set, the mode is **75**, which occurs 3 times.

6. **C**

To find the standard deviation for the entire project we cannot just add the deviations of the individual tasks. The correct way of doing it is as follows:

$$\text{Variance of individual tasks} = (\text{std. deviation})^2$$

$$\text{Variance (Task A)} = 4 \text{ weeks}$$

$$\text{Variance (Task B)} = 9 \text{ weeks}$$

$$\text{Variance (Task C)} = 36 \text{ weeks}$$

$$\text{Total project variance} = 4 + 9 + 36 = 49$$

$$\text{Total project standard deviation} = \sqrt{\text{total variance}}$$

$$= \textbf{7 weeks}$$

Venn Diagrams

WHAT IS A VENN DIAGRAM?

Introduced by the British logician John Venn in 1881, a Venn diagram represents mathematical and logical relationships between different groups of data. Simple Venn diagrams consist of circles representing the data sets. The circles can be overlapping or disjoint (separate), depending on how the data sets are related. A Venn diagram shows at a glance, in a very visual way, the commonality and relationship between the data sets. Venn diagrams can be drawn using shapes other than circles, too. Figures A.1 and A.2 illustrate two simple Venn diagrams.

The Venn diagram shown in Fig. A.1 depicts a situation in which there is an overlap between the schedule risks and cost risks of a project. The overlapping area, called the intersection of the two data sets, depicts the set of common risks that impact both schedule and cost. For you as a Project Manager, those common risks might be the ones that are most important to monitor closely.

The relationship between the data sets in this Venn diagram is represented in mathematical terms as follows:

$$\text{Common risk} = (\text{schedule risks}) \cap (\text{cost risks})$$

where $\cap$ is the "intersection" operator.

$$\text{Project risk} = [(\text{schedule risks}) \cup (\text{cost risks})] - [(\text{schedule risks}) \cap (\text{cost risks})]$$

where $\cup$ is the "union" operator.

The Venn diagram shown, in Fig. A.2 illustrates a situation in which there is nothing in common between the set of schedule risks and the set of cost risks affecting a project. The two disjoint (separate) data sets are represented by circles that do not overlap. In this case, the combined project risk is the "union" of the two data sets (i.e., schedule risks plus cost risks).

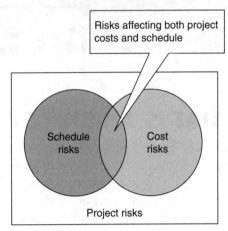

Risks affecting both project costs and schedule

Schedule risks

Cost risks

Project risks

Figure A.1 Venn diagram: Overlapping project risks.

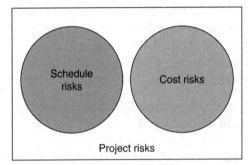

Schedule risks

Cost risks

Project risks

Figure A.2 Venn diagram: Distinct project risks.

In mathematical terms:

$$\text{Common risk} = 0$$

because there is no overlap in the Venn diagram.

$$\text{Project risk} = (\text{schedule risks}) \cup (\text{cost risks})$$

where $\cup$ is the "union" operator.

PERT Analysis Using Microsoft Project

Microsoft Project (MS Project; 2003 and 2007) provides a PERT Analysis tool (toolbar) to develop and analyze schedules and performs quantitative analysis of project risks. This tool allows the Project Manager to estimate project and task durations and also to assess the impact of several risks on the project if they were to occur.

Before you begin PERT Analysis, you need to complete the following tasks:

- Define the project in MS Project.
- Set up the project's working times (days and times).
- Define the holiday calendar.
- List the resources (people) working on the project.
- List the tasks in the project and assign them to resources.

A sample project with some specific tasks as defined in MS Project is shown in Fig. B.1.

Once the project and its task and attributes have been set up, you can start the process of using PERT to analyze the project's/task's estimated durations. Start off by activating the PERT Analysis toolbar in MS Project. Select the View menu in MS Project and then select PERT Analysis from the Toolbar menu, as indicated in the screen shot in Fig. B.2.

Once the PERT toolbar has been selected, it will appear as shown in Fig. B.3 (outlined in black on the left side of the screen).

The PERT Analysis toolbar consists of seven buttons, each serving a different purpose. The functions of these buttons in this toolbar are described below:

1. **Optimistic Gantt** The purpose of this button is to specify the optimistic estimate for the completion of a task.
2. **Expected Gantt** The purpose of this button is to specify the most likely and expected estimate for the completion of a task.
3. **Pessimistic Gantt** The purpose of this button is to specify the pessimistic estimate for the completion of a task.

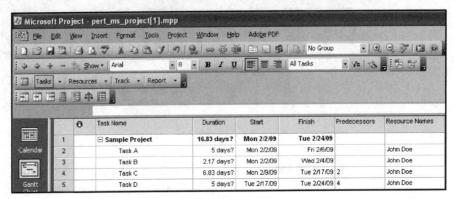

Figure B.1 Microsoft Project: sample project.

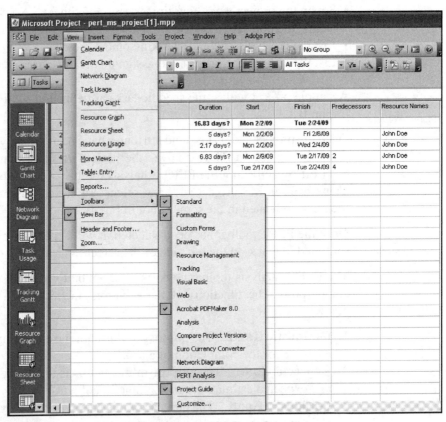

Figure B.2 Starting PERT Analysis in MS Project.

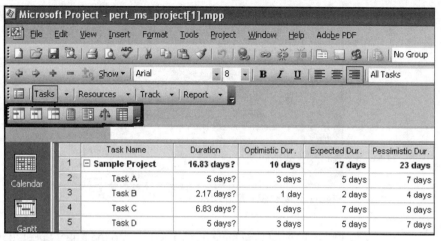

Figure B.3 PERT toolbar.

4. **Calculate PERT** The purpose of this button is to use the specified optimistic, most likely, and pessimistic task duration estimates to calculate the expected task duration using the PERT formula:

$$\text{Expected duration} = [P + (4 \times M) + O] \div 6$$

5. **PERT Entry Form** The PERT Entry Form button is an alternate way to specify the optimistic, most likely, and pessimistic durations for a selected task. It can be used instead of using the method listed in (1), (2), and (3) above.

   Please note that the task should be selected prior to launching the Entry Form. Multiple tasks can also be selected prior to launching the Entry Form. In the case of multiple tasks, the estimates you define will apply to all the selected tasks.

6. **Set PERT Weights** By default, the PERT Weights are set to 1 for the pessimistic estimate, 4 for the most likely estimate, and 1 for the optimistic estimate. This is inline with the assumption that the distribution is "uniform."

   The default PERT Weights can be changed as needed, depending on the situation.

7. **PERT Entry Sheet** The PERT Entry Sheet provides a worksheet for documenting the estimates for each task on the project. It is similar to the PERT Entry Form; the difference is that the Entry Sheet provides a way to define estimates for all tasks on the same screen.

Given in Fig. B.4 is a screenshot of the PERT Entry Sheet, with estimates provided for each task in the sample project.

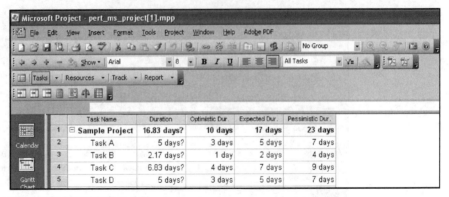

Figure B.4 PERT Entry Sheet.

Once relevant data have been inserted into the Entry Sheet, it is time to put the PERT wizard to work and calculate the expected durations for the individual tasks and the project as a whole, too. Clicking on the "Calculate PERT" button [as discussed in point (4) in the preceding list] in the PERT Analysis toolbar triggers this calculation.

The result is the calculated duration as illustrated in the second column of the screenshot in Fig. B.5.

As shown, the PERT calculation indicates that the project duration is expected to be 16.83 days.

This number will vary under the following circumstances:

1. Relationships between the tasks change.
2. Optimistic, most likely, or pessimistic estimates change.
3. Resource availability changes.
4. PERT Weights are changed from the default.

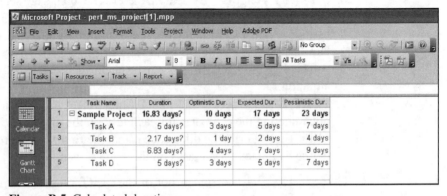

Figure B.5 Calculated duration.

Index

A

Accelerated depreciation, 15
Actual Cost (AC), 157
Actual cost of work performed (ACWP), 157
Analogous estimating, 77
Arithmetic mean, 202
Arrow diagramming method, 88
Atypical variance, 168
Average project score, 5

B

Bar charts, 190
Bell curve, 180
Benefit-cost ratio, 13
Benefits measurement, 3
Beta distribution, 179
Bottom-up estimating, 77
Brainstorming, 6
Break-even analysis, 18
Break-even point, 27
Budget at Completion (BAC), 167
Budgeted cost of work performed
    (BCWP), 157
Budgeted cost of work scheduled
    (BCWS), 157

C

Cause-and-effect diagram, 191
Central tendency, 202
Check sheet, 195
Communication channels, 40, 147
Constrained optimization, 7
Contingency reserve, 79
Contract cost estimation, 124
    cost reimbursable, 125
    fixed price, 124
    time and materials, 125
Contract costs, 126
Control charts, 188
Cost baseline, 79
    contingency reserve, 79
Cost budget, 79
    management reserve, 79
Cost Performance Index (CPI), 156–158
Cost reimbursable, 125
Cost reimbursable contract, 125, 139
Cost Variance (CV), 156–157
Cost-benefit analysis, 6

Crashing, 40, 101
    fast tracking, 101
Critical path, 40, 86
    arrow diagramming method, 88
    Critical Path Method (CPM), 89
    graphic evaluation and review technique
        (GERT), 89
    network diagramming methods, 87
    precedence diagramming method, 87
Critical Path Method (CPM), 89

D

Decision tree, 40–65
Delphi method, 6
Dependent events, 44
Depreciation, 14
    declining balance method, 15
    straight-line depreciation, 14
    sum of the years method, 15
Direct costs, 79
Dispersion, 202
Dynamic programming, 7

E

Earned Value (EV), 157
Earned Value Analysis, 167
    assessment, 156
        Actual Cost (AC), 157
        budgeted cost of work performed (BCWP),
            157
        budgeted cost of work scheduled
            (BCWS), 157
        Cost Performance Index (CPI),
            156–158
        Cost Variance (CV), 156
        Earned Value (EV), 157
        Planned Value (PV), 157
        Schedule Performance Index (SPI),
            156–158
        Schedule Variance (SV), 158
    forecast, 167
        Budget at Completion (BAC), 167
        Estimate at Completion (EAC), 168
        Estimate to Complete (ETC), 167
        Variance at Completion (VAC), 169
Economic models, 3
Estimate at Completion (EAC), 168
Estimate to Complete (ETC), 167

F

Fast tracking, 101
Fishbone diagram, 191
Fixed costs, 78
Fixed price, 124
Float, 86
    free float, 87
    negative float, 87
    project float, 87
    total float, 87
Flow chart, 194

G

Gantt charts, 213
Graphic Evaluation and Review Technique
    (GERT), 89

H

Histograms, 190

I

Incentive, 126
Independent events, 44
Indirect costs, 79
Inspection, 194
Integer programming, 7
Internal rate of return (IRR),
    3
Intersection, 211
Ishikawa diagram, 191

L

Lease-or-buy analysis, 26
Linear programming, 7

M

Make-or-buy analysis, 26
Management reserve, 79
Mean, 182, 202
Median, 203
Microsoft Project, 213
Mode, 203
Mutually exclusive events,
    43

N

Negative float, 87
Net present value (NPV),
    3–4
Network diagramming methods,
    87
Normal distribution, 180

O

Opportunity cost, 13
Optimistic Gantt, 213
Outliers, 189
Overhead fee, 126

P

Pareto charts, 190
Payback period, 3
Program Evaluation and Review Technique
    (PERT), 40, 89, 114, 213
Pessimistic Gantt, 213
Planned Value (PV), 157
Point of Total Assumption (PTA), 40, 138
Precedence diagramming method, 87
Present value, 3–8
Probability, 40
    probability distribution, 178
        bell curve, 180
        beta distribution, 179
        normal distribution, 180
        triangular distribution, 179
        uniform distribution, 180
    probability fundamentals, 41
        dependent events, 44
        independent events, 44
        mutually exclusive events, 43
Project charter, 1
Project cost estimation, 40, 76
    analogous estimating, 77
    bottom-up estimating, 77
    parametric estimating, 78
Project cost pyramid, 77–78
Project float, 87
Project initiation, 1
Project selection models, 2
    benefits measurement, 3
    constrained optimization, 7

Q

Quality control tools, 178, 188
    bar charts, 190
    cause-and-effect diagrams, 191
    check sheets, 195
    control charts, 188
        rule of seven, 189
        special cause, 189
    fishbone diagrams, 191
    flow charts, 194
    histograms, 190
    inspection, 194
    Ishikawa diagrams, 191
    outliers, 189
    Pareto charts, 190
    run charts, 192
    scatter diagrams, 193

R

Range, 205
Return on assets (ROA), 16
Return on investment (ROI), 17
Return on sales (ROS), 16
Review board, 6
Rule of seven, 189
Run charts, 192

S

Scatter diagrams, 193
Schedule Performance Index (SPI), 156–158
Schedule Variance (SV), 158
Scoring models, 4
Shewhart, 182
Six Sigma, 181
Special cause, 189
Standard deviation, 114, 182, 204–205
Statistical concepts, 201
    central tendency, 202
    mean, 182, 202
    median, 203
    mode, 203
    standard deviation, 114, 182, 204–205
    variance, 114, 182, 205
Statistical sampling, 193
Straight-line depreciation, 14

Sum of the years method, 15
Sunk costs, 14

T

Time and materials, 125
Total float, 87
Trend chart, 192
Triangular distribution, 179
Typical variance, 168

U

Uniform distribution, 180
Union, 211

V

Variable costs, 79
Variance, 114, 182, 205
Variance at Completion (VAC), 169
Venn diagrams, 41, 211
    intersection, 211
    union, 211

W

Working capital ratio, 17

ABOUT THE AUTHORS

Vidya Subramanian, PMP   Vidya is an experienced PMP with professional experience in different roles spanning all phases of a software development life cycle. She currently works at a leading financial institution in Virginia. Her core competencies include process improvements, business analysis, and project management.

She holds a Bachelor's in Accounting and a Master's in Computer Software Applications from Mumbai, India. She also holds a Master's in Information Systems from Virginia Tech. She is an active member of the Project Management Institute and a certified PMP and Scrum Master.

Ravi Ramachandran, PMP   Ravi is a professional information technology consultant in Virginia, experienced in technology consulting, managing, analyzing, and implementing software solutions for clients. His expertise in the technology sector spans various industries and his core competencies include client relations, business analysis, technical consulting, and off-site management.

Ravi holds a Master's in Computer Science from the University of Texas at Dallas and an MBA from Virginia Tech. He is an active member of the Project Management Institute and a certified PMP.